At Least Know This

At Least Know This

Essential Science to Enhance Your Life

GUY P. HARRISON

 Prometheus Books

**59 John Glenn Drive
Amherst, New York 14228**

Inquiries should be addressed to
Prometheus Books
59 John Glenn Drive
Amherst, New York 14228
VOICE: 716–691–0133 • FAX: 716–691–0137
WWW.PROMETHEUSBOOKS.COM

22 21 20 19 18   5 4 3 2 1

Library of Congress Cataloging-in-Publication Data Pending

Printed in the United States of America

[DEDICATION TO BE ADDED AT A LATER DATE]

CONTENTS

INTRODUCTION

Humankind can never know itself or come close to its intellectual and creative potential so long as most members of the species continue to suffer critical gaps in fundamental knowledge. Factually inaccurate conclusions, irrational beliefs, and allegiance to extraordinary claims that are demonstrably false plague us at every turn. As a species we labor under a crippling weight of willful ignorance. How can any society expect to make wise decisions regarding complex challenges such as climate change, biodiversity loss, possible viral pandemics, genetic engineering, and energy needs when many if not most people do not even know how old the universe and Earth are, where humans came from, or how life works? For example, at a time when drug-resistant germs are becoming more common, 45 percent of American adults don't understand that antibiotics work on some species of bacteria but not on viruses.[1] A similar ratio of Europeans, 46 percent, know this, while 28 percent of Japanese people and 18 percent of Russians do. Empowered with the scientific process, humankind has made many spectacular and invaluable discoveries over the last few centuries. Is it not tragic that all or at least most people are not fully aware and applying as much of this knowledge as possible to benefit their lives?

The United States is one of the world's leaders in science education[2] and yet less than forty percent of American adults the possess the minimal science knowledge needed to read and understand the *New York Times* science section.[3] One survey found that only about forty-four percent of Americans could give a brief correct explanation of what DNA is and eighty percent did not know what a stem cells are.[4] Virtually all physicists and astronomers understand that the universe began about 13.8 billion years ago with the Big Bang, but only twenty-one percent of the American public agree that it did.[5] A mere twenty-seven percent accept that the Earth is 4.5 billion years old.[6] All of this matters because evidence-based knowledge impacts if not determines the success of our actions, the range of our vision, and our ability to appreciate existence.

Humankind is condemned to navigate in partial darkness if we do not base important decisions on the best-available facts and theories. The purpose of this book is to focus readers on the simple questions with profound answers. These are the queries that have great relevance to our lives and worldview. Answering them to some degree, from the scientific perspective, can improve the quality, effectiveness, and efficiency of our ideas and actions. This is the necessary knowledge, the invaluable information that can enhance the life of an individual and raise the overall quality of our societies.

My goal is to leave readers not only more informed about who they are, where they come from, what is going on around them, and so on, but also to *feel* differently about themselves, the life they share the Earth with, and the entire universe. We tend to place so much emphasis on immediate and practical uses of knowledge—pass a test, build something, perform a task—that we overlook the positive emotional power of fundamental knowledge about the reality we find ourselves in. It is common for scientists and rationalists to warn people about the danger of emotions clouding thought processes and derailing our ability to reason well. I certainly have done this in some of my previous books. But let's not make the mistake of condemning all emotion all the time. Learning key facts and concepts about ourselves and our surroundings can stimulate and inspire. Don't fight it. Get excited. Let some emotions soar as you read this book. Allow yourself the freedom to find more passion for life in the scientific knowledge presented here. New awareness of who we are and what is going on around us can leave us feeling more alive and more awake as human beings.

This is not a book of science trivia or fringe facts for self-amusement. The pages ahead are not conversation fodder meant to impress friends. This is the important stuff. This information is the foundation from which further learning rises. The topics covered here matter to your existence as a conscious and highly intelligent life-form in the twenty-first century. This book is a chance to catch up and fill in gaps that anyone may have in his or her understanding about the most important things, events, and processes of all.

No one should feel embarrassed about having forgotten or never learned who are our distant ancestors are, when the universe and

Earth were formed, how atoms work, why microbes matter, and so on. While striving to make good grades, land a job, achieve social acceptance and find a mate, many smart and hard-working people rush right by this fundamental knowledge. Ignorance on these matters does not necessarily have anything to do with one's intelligence or work ethic. So let's not waste time blaming parents, schools, or society in general for things we do not know. Let's get busy learning instead. I hope you enjoy exploring the essential science within this book. May it inspire and enhance your life.

Guy P. Harrison,
Earth

Chapter 1

THE *WHO, WHAT, WHY, WHEN, WHERE, AND HOW* OF EVERYTHING

W*ho are we? Where are we? How did we get here? How long have we been around? How old is life? What is everything made of? When did everything begin? How do brains work? What lifeforms share your body with you? How will everything end?* These are simple questions, the kind a child might ask, but most adults would struggle to come up with substantial, evidence-based replies to them. We all can know competent answers to these and other such questions, however, thanks to the dedicated work of numerous scientists over many years. Though the current answers may be incomplete, inconvenient, or even disorienting for some, they are invaluable to us and should be grappled with by all.

Addressing these questions gets us to primary knowledge that offers surer footing amidst so much complexity and change. These are the inquiries that open doors to the most crucial discoveries, facts, and ideas. These answers matter in a practical sense because they help us become more sensible, safer, efficient, and productive. They also feed a human quality that we often forget we have—curiosity, the desire to explore, the need to know about our world and universe, as well as the objects and processes that make up our world and universe. They pack emotional power as well. The satisfaction of discovery and unpacking the inner workings of our brain and mind opens up new possibilities for learning more about ourselves and feeding our curiosity. This is the information that can inspire us to stand taller and enjoy a deeper appreciation for our existence as intelligent life.

Utilizing logic, reason, and evidence to determine, or intelligently approximate, how old the universe, Earth, and life are and how our world might end one day can make our daily experiences richer and our lives fuller because our personal time in this world is not infinite and recognizing our own mortality can instill a desire to learn as much as we can in the short time we have. You want this knowledge. Knowing more enables you to see more, to reach further and dream larger. It is one thing to glance at pretty little lights in the night sky, but quite another to stare up at the universe, understand your connection to it, and feel the weight of its size, forces, movements, and grandeur. To stand before a mirror and find but one person looking back at you is a lost opportunity for amazement and inspiration. Better to know the scientific origin story of your species so that you might see in your own eyes faint reflections of the perilous and improbable odyssey of life that led to you.

Hiking through a thick, green forest may be enjoyable for everyone to some degree. But every step is a jolt of excitement when taken in the light of scientific awareness. Once you understand your kinship to all life, a mere walk among trees takes on the feel of an epic voyage. Look down. Beneath your feet, in a single gram of soil is a thriving galaxy of life, home to many millions of individual microbes.[7] Look up. Find the tallest nearby tree. There you will see a bustling megalopolis, its leaves, branches, trunk, and roots are the homes and workplaces of countless residents and transients, including birds, mammals, reptiles, insects, arachnids, other plant species, fungi, bacteria, and viruses. Every tree stands as a spectacular monument to life.

Ideas, dreams, and goals begin with knowledge. The woman who understands something about the evolution of a human brain may be more appreciative and thoughtful of how she goes about using and caring for her own. The man with the benefit of at least a vague notion of the astounding stretch of time that life has existed on this planet is better positioned to recognize the beauty, wonder, and value of biodiversity. He also might find greater meaning in the joys and opportunities that his own relatively brief life can offer. The child who has learned something about the omnipresent swirling sea of microscopic life is never far from excitement and inspiration. She or he may look in any direction at any moment and sense the presence of invisible lifeforms both beautiful and monstrous.

What is life? Why does is change over time? How are we connected to other life? How much life is on Earth? These are profoundly important questions for all of us and our ability to answer them can determine the wisdom of our decisions and actions. What might happen if every politician and CEO knew that an overwhelming majority of life on Earth today has not even been identified or named, much less studied? According to estimates based on scientific surveys and analysis, it is likely that as much as 86 percent of living land species and 91 percent of living ocean species still await discovery and description.[8] Would our species clear, burn, and loot wild habitats at current rates if at least most people understood the potential economic value of so much undiscovered life? Would the public tolerate the destruction of nature at current levels if it were commonly known how long it took present biodiversity to evolve and how long it might take ancient ecosystems to recover from human devastation, if ever? It must be immensely difficult, if not impossible, for any business person, politician, or voter to consistently think rationally and act wisely about virtually anything of consequence in the absence of a basic, big-picture, and fact-based perspective from which to begin.

Current ignorance about so much of the life beside us can serve as inspiration for us to pick up the pace of discovery. The fact that science can only provide us with an incomplete picture can motivate everyone to learn as much as we possibly can about what knowledge we do have. Robert M. May, an Oxford University ecologist and former Royal Society president, describes our problem:

> It is a remarkable testament to humanity's narcissism that we know the number of books in the US Library of Congress . . . but cannot tell you—to within an order of magnitude—how many distinct species of plants and animals we share our world with. . . . We are astonishingly ignorant about how many species are alive on earth today, and even more ignorant about how many we can lose yet still maintain ecosystem services that humanity ultimately depends upon.

John Cryan is a principal investigator at the APC Microbiome Institute in Ireland and has been named one of the "World's most Influential Scientific Minds." He suggested to me that is impossible to know our planet or ourselves without knowing something about what microbes are up to. "We are living in a microbial world," he said. "They were there first and we have co-evolved with them. There never has been a time when they haven't played a role in programming our systems. In terms of genes we are more than 99% microbial and have more microbial cells than human cells."[9]

All around us there is life, big and small, doing fascinating and important things. Geological processes are playing out beneath our feet every moment that have profound impacts on our lives. And above us there is a universe of wonder that is waiting to excite every human mind.

SIX QUESTIONS

Student journalists traditionally are taught to answer six key questions within the news articles they write.[10] If the final published piece doesn't address the *who, what, why, when, where, and how* of a topic or event there could be problems. It doesn't matter if a journalist is writing about an election result, a sports game, or a battle; addressing those questions is critical. Failing to satisfy this list means leaving holes in the story that may confuse or frustrate readers. It shortchanges them on the most basic and relevant information. What many people don't realize, however, is that these same six questions are just as important outside the walls of newsrooms and journalism classes.

When we neglect to answer *who, what, why, when, where,* and *how* about our species, our environment, our past, and so on, we shortchange ourselves in life. It means there will be significant holes in our self-perception and worldview. Finding sensible, realistic context is elusive when basic information is missing.

MIND THE GAP

If you are not convinced that the six questions have any importance to you, consider them in a more personal framework. Ignorance of the *who, what, why, when, where,* and *how* in the context of your particular life would mean not knowing where you were born or not remembering where you grew up. Imagine not knowing where you are now or how you got there. How disorienting and scary would it be to have little or no awareness of not only your personal past, but no knowledge of the history and customs of the society you live in? How would it feel to be unsure if the people you come into daily contact with are close family members or strangers so alien that they may as well be from another world? What would it mean to have little or no sense of where your house is in relation to the city, country, and world? Sounds a bit like being trapped inside a *Twilight Zone* episode with no ID, library card, or Wi-Fi access. It's an odd thing to consider, but most people live in a similar state of disorientation. Right now, in the twenty-first century, with so many hard-earned scientific discoveries and so many real answers available, we remain, overall, a lost species, floundering in a sea of unnecessary ignorance. Broad-based learning and acceptance of the most basic aspects of our existence is likely the only way we can ever wake up and grow up as human beings.

FAR HORIZONS

To make the most of the information contained within this book it will be necessary for readers to repeatedly assess how and where they fit into their world. It is easy for us to settle into intellectual safe-zones built from our hopes, fears, and traditions. Part of the standard human experience is believing in false walls and other cognitive limitations promoted by whatever culture we happen to have been born into. To a degree, this can be benign, functional, or positive. Some of it, however, may be socially corrosive and stand in direct opposition to an evidence-based or scientific worldview. When we fall too hard for artificial divisions and manufactured labels, it can be difficult if not impossible to bring reality and the bigger picture into focus. Therefore, it is important to recognize that we are more than citizens of one country, or members of one political, religious, or racial group. No one should cringe or run away from any of this. I am not asking readers to abandon affiliations they enjoy or feel are of vital importance to them, only to see and acknowledge what lay beyond the fenced-in confines of a typical human life.

Most people throughout most of prehistory and history, have lived as fractured, tribal beings in psychological isolation from the rest of humanity. Home has been defined as one small natural habitat or region, a village, neighborhood, city, nation or empire. Through science, however, we understand that reality is larger than this. We are a young and very closely related species. Our home is more than a mailing address. There is more to us than flesh and blood. We fit into a vast and fascinating system of atoms, molecules, lifeforms, planets, stars, galaxies, and super galaxies. Our home is the universe. Our existence is more astonishing and wonderful than most of us realize.

We all can know, if we wish to, that humankind far exceeds the sum of divorced and disparate clans. We are as large as our knowledge. Our real horizons stretch as far as whatever limits the universe may set. Has there ever been a more exciting time to be a human? So much accumulated knowledge is available now and more learned each moment. We grow every day as we see and reach further. We have explored the natural world, near and far, and discovered much about ourselves. But not everyone has noticed.

Many people have opposed the march of scientific discovery because they fear it deflates and reduces humanity, somehow diminishing our stature as special beings. Copernicus and Galileo pushed Earth away from the center of the solar system and universe. Darwin moved us from the center of all life. And then Freud made us wonder if consciousness was even at the center of our own mind. But fears of our demise proved unfounded. We did not shrink under the bright lights of scientific scrutiny. The opposite has happened. Learning about ourselves, our neighbors, and our home planet makes us more special, not less. Answers from science expand and enrich the human definition. Knowledge adds value to our humanity. Even if it were to turn out that intelligence is common throughout the universe and there is nothing so rare or special about us after all, I still would argue that there is plenty to be proud of. Faced with incomprehensible complexity and mystery all around us at least some people decided to be more than passive, incoherent bystanders. Good scientists and good thinkers chose to be active explorers and learners, to *participate* in this spectacular universe. Fortunately, anyone anywhere can join this trend.

What are we? To which domain, kingdom, phylum, class, order, family, and genus do humans belong? The answers have relevance beyond a high school biology quiz because they help to inform us about our relationships and kinships with all the other lifeforms with which we share this planet. *How old are anatomically modern humans? How did prehistoric people live and what does that mean for us today?* A person who is uninterested in being able to answer these questions cannot expect to know herself or himself well. Missing knowledge that relates to our worldview and self-perception will be filled in with potentially inaccurate information, or made-up stories, or simply left blank. None of these are good options because when separated from basic realities people are left less able to think rationally, make good decisions, and identify dangerous ideas. For example, enduring problems such as racism, runaway nationalism, and xenophobia depend on specific gaps in information if they are to thrive and do their worst. It is all too easy for otherwise good and smart people to fear and hate "them" when they don't know who humans are, how we got here, and where we fit in with the *Homo* genus and all the rest of life. It is easier to divide, fear, hate, and sometimes kill our sisters and brothers when we define ourselves based primarily on mistakes and fantasy. Racism, genocide, and other such self-imposed disasters may not be just likely in the absence of scientific self-awareness, they may be inevitable.

Where do you live? How old is your home? What is your home like? Most of us know our street address. But how many know the age and general orientation of our ultimate human home—Earth, our solar system, and the universe? Surveys suggest that many millions of people have little or no idea about any of this.[11] It is impossible to possess a realistic understanding of how our home planet works, how it has changed and continues to change, without first being aware of its immense age and relationship to other celestial bodies and systems. There could hardly be a simpler and more straightforward question than: *Where am I?* Many people, however, struggle to place their planet in anything close to correct spatial context with the Sun, other planets in our solar system, our galaxy, and the known universe. Surveys have shown, for example, that only 74 percent of Americans and 66 percent of Europeans understand that our Earth revolves around the Sun rather than vice versa.[12] Twenty-seven percent of American adults don't know the difference between the science of astronomy and the pseudoscience of astrology.[13]

Donald Lowe, a professor of geological sciences at Stanford University, believes that knowing basic facts and concepts about our planet is essential to knowing ourselves, our civilization, and something about our collective path ahead. "I think that an educated public is, or should be, a curious public and some of the things that we should want to know include the history and origin of our planet," Lowe told me. "That origin and history bear strongly on who we are and where we ultimately came from. Part of being an educated public is having that knowledge available, even if many choose to spend their learning efforts elsewhere. More practically, the bulk of the resources upon which we depend come from the Earth. oil, gas, gold, iron for the steel in our cars and buildings, copper, gypsum for the wallboard in our homes, rare earths which are so much in the news these days . . . even lowly sand and gravel for roads, all come from the Earth. How can we responsibly develop and use these resources unless we know where they come from and how they form? These resources have determined the paths of human history and the fates of empires."[14]

Where you live is not a trivial piece of information. Knowing your place, in time and space, in a universe as big and awesome as ours is precisely the kind of basic knowledge that can expand thoughts and spark boundless ambitions over a lifetime. But twenty-eight percent of Americans do not even understand that a light year is a measure of distance.[15] Missing that, how can they begin to grasp their location? Anyone who lacks a minimal awareness of neighboring planets, the structure of our galaxy, and a certain gigantic fireball—that thing that lights and warms the Earth—might be sleeping through life. Those who can should elevate their awareness to a basic level of intellectual competency as a modern human. There are few excuses available here. Short of being mired in extreme poverty, rendered incompetent for some reason, or trapped inside a closed society like North Korea, it's just not that difficult to learn the basics of your existence. This observation is not meant to insult anyone but rather to encourage everyone.

How will we survive? This is an important question, one every human ought to ponder, research, and try to answer. Yet few people are aware of what science says about the survival of humankind. Thanks to a few movies and TV shows, most people probably know now that a strike by a ten-mile-wide asteroid would put an end to our world as we know it. But what about supervolcanoes? The threat they pose to civilization and human existence is far from common knowledge. Very few people know, for example, that the Toba eruption around 73,000 years ago in present-day Sumatra sent so much smoke, soot, and dust into the atmosphere that humankind might have come close to extinction in the immediate years thereafter. Had that global dimming and cooling event been any more severe our species may have starved and faded away forever.

You might think that existential risks—possible events severe enough to cause our extinction—would be of obvious interest to everyone. Strangely, however, many people seem to view it as a topic unworthy of much concern or study. Governments certainly don't take them very seriously based on the relative investments made. As recently as 2010 NASA was budgeted a mere $4 million per year to find dangerous asteroids and develop means to defend our planet and humankind against them. That rose to $50 million in 2016.[16] Given the stakes of possible extinction and compared with money spent elsewhere, however, it's still nearly nothing. One aircraft carrier, for example, cost the US government $13 billion[17] and Americans spent more than $16 billion in 2016 alone on Botox wrinkle treatments, nose jobs, liposuction, and other plastic surgeries.[18] But closing our eyes and maintaining a carefree attitude won't protect us from real, evidence-based potential catastrophes. Ignorance, intellectual and financial neglect, abandoning responsibility—whatever we call it—does not seem like a wise survival strategy for the Earth's smartest lifeform. When we learn about space, our planet, our species, and all life we see that the future is uncertain and bad things happen around us all the time. Anyone who understands the basics of evolution and extinction, and knows something of the natural history of this planet will be aware that lifeforms hit dead-ends. It is a normal outcome. Extinction is part of nature's standard operating procedure and there is no reason to think we are immune to this harsh and final punctuation. Many biologists estimate that extinction has already been the fate of *ninety-nine percent or more* of all species that have ever existed on Earth.[19] This estimate is mentioned frequently these days due to environmental concerns but how many of us pause and contemplate the deeper meaning of it? *Ninety-nine percent or more.* Given enough time, most roads are dead-ends. Given enough time, maybe every road is a dead-end. Our advantage is that we are the first species ever on this planet capable of contemplating and acting on distant possibilities. We have the mental power to imagine and shape the futures we desire.

Knowing who we are is a critical aspect of our survival. Self-awareness, based on the best current evidence, means understanding that we are a relatively young and in many ways vulnerable species. Some existential threats—nuclear war is one—come from within; others—supervolcanoes and asteroids are two—come from without. Some might be a combination of human intelligence and indifferent nature: a deadly virus made worse in a lab, for example. A viral plague might emerge from a human lab or naturally. Recognizing and reacting appropriately to the biggest possible dangers, from a perspective based on science and reason, is necessary to allow us the best chance of long-term survival. And it begins with a simple but important question: *How will we survive?*

WHAT IS SCIENCE?

Science is the name given to one of humankind's greatest cognitive innovations and its most important revolution. This way of thinking and interacting with reality has radically changed us and will continue to do so. Science is the means by which we escaped so many of nature's shackles. It predates the Age of Enlightenment, Galileo Galilei, Isaac Newton, and even Aristotle. Prehistoric humans were doing crude but effective science more than a million years ago, for example, when they observed fire, hypothesized about its properties, experimented, drew conclusions, developed some workable theory of fire, and tamed it. It was a rudimentary, informal, and undeveloped version of the scientific method that helped our distant ancestors craft more effective tools, build better shelters, and invent seaworthy rafts. On the strength of something close to scientific thinking, multiple human species were able to spread and flourish over the last two million years.

Critical thinking and the wise application of knowledge enable us to reach beyond the limits of instinct and transcend many of our evolutionary obstacles. Science, done well, is a precious and remarkably effective tool, a system of discovery and confirmation that relies on evidence and honesty rather than hope and loyalty. It also is the source of an immense and crucial body of knowledge, arguably our greatest treasure. Moreover, as a way of reasoning that is accessible to everyone, science empowers us to see around misperceptions and think through confusion to avoid delusions.

Science offers us the potential, or the possibility, to make quicker progress. By far, most of the truly important information about ourselves, our world, and the universe was discovered and brought into focus with the scientific method or with evidence-based thinking that may be described as scientific. It is our great failure that so many of us ignore or underappreciate this wealth of knowledge and know-how. It is to our collective peril that most people consistently fail to apply the scientific method or use scientific thinking toward positive ends for themselves, their families, and their societies.

Think for a moment about how much we have learned and how far we have come. We progressed from foot locomotion and stick and stone tools to megacities and nuclear-powered submarines. The pace has often been astonishing. Orville and Wilbur Wright, for example, achieved powered flight at Kitty Hawk, North Carolina in 1903. A mere 66 years later, humans were conducting science experiments while standing on the surface the Moon. The volume of our accumulated scientific answers in the early twenty-first century is enormous. It spans quarks, perhaps the tiniest sub-atomic particles of all, and the Hercules-Corona Borealis Great Wall, perhaps the largest "thing" in the universe[20]. To date, an impressive amount of knowledge has been harvested and stored. And all should feel a sense of ownership. Science, as a way of thinking, is a human endeavor. Therefore, it is your birthright as much as anyone else's. No one person can consume and know all of science, of course, but we all can wade into enough of it to gain practical awareness and a useful understanding of who we are, what we are doing here, and where we might go.

BILLIONS IN THE DARK?

Significant numbers of people around the world are unfamiliar with basic knowledge about the basic structure of, well, *everything*. In 2016 the National Science Board[21] published the findings of several studies, revealing that the slimmest majority of American adults, 51 percent, understand that an electron (a part of an atom) is smaller than an atom. It's worse in Japan and China where only 28 percent and 27 percent, respectively, know this. Among Europeans and South Koreans, 46 percent understand this size ratio. Russians get it right at a rate of 44 percent, Malaysians 35 percent, and Indians 30 percent. Canadians rate the highest of all countries surveyed with 58 percent aware that electrons are smaller than the atoms they help to make up; still a dismal result. These surveys suggest that two-thirds or more of the human population is likely in the dark about the most elementary and important features of reality.

Meanwhile, most American adults, 52 percent, believe haunted houses are a real thing and thirty-five percent think technologically advanced extraterrestrials visited the Earth during ancient times. Nineteen percent believe that fortune tellers and psychics can foresee the future and 16 percent are convinced that Bigfoot is still out there, somewhere in the Pacific Northwest eluding capture and confirmation after all these years.[22] For the record, science offers nothing substantial to back up any of this. These claims and others like them could be true, of course. But in the absence of scientific confirmation to a respectable degree, why pretend to know that they are? A big problem with unscientific thinking—the failure to be skeptical and demand credible evidence—is that it leaves us open to invasion by virtually any belief, no matter how ludicrous or unlikely. Critical thinking, skepticism, and a scientific worldview are necessary to keeping our minds from being overrun by bogus beliefs, lies, and common mistakes of reason and perception.

Current scientific facts are not permanent truths. At least we can't think of them that way. All concepts, conclusions, and theories remain malleable, potentially under construction forever as new knowledge is uncovered. But this doesn't mean they don't have great value or are useful. Our best present scientific knowledge offers us a very helpful view of basic and functional reality now.

Much of this information is so ultra-complex as to be inaccessible to laypersons. Quantum entanglement, for example, is maddeningly bizarre and contradictory concept in physics. And the best possible explanations for the origin of life that science can give us today are frustrating and unsatisfying because they are mostly speculation. Some of the more exciting ideas in physics and astronomy come wrapped inside thick blankets of mathematics—a pure horror to some. But no one should be intimidated by complexity put off by uncertainty. With minimal effort, anyone can learn a great deal that is useful and stimulating. I once gave a fast-paced lecture on the Big Bang, black holes, and Dark Energy to a classroom of twelve-year-old students, all of it jammed into a single hour. They leaned into the knowledge and considered the scientific discoveries and mysteries I presented to them without fear or reservation. If anyone was intimidated that day it was me—a non-astrophysicist— because at the end of my talk I had to field a relentless barrage of great questions from excited kids. If children can be enthusiastic and curious about living in a complicated, weird, beautiful, scary, and strange universe, why can't every adult? The responsibilities and stress of adulthood are no excuse. If anything, learning more about life, humankind, our world, and the universe offers some inspiration to help us through the daily grind of modernity. To miss, ignore, or deny this core knowledge of human existence is to live a diminished life. *Who, what, why, when, where, and how* are questions that should stalk us—in a good way—throughout our lives. There's also that Socratic quote about the unexamined life not being worth living to consider.

IMPERFECT AND INVALUABLE

It is important to be clear about what science is and what it is not. Given its remarkable and consistent productivity, it may be easy for some to become so enamored as to have blind faith in it. This would be a mistake. Modern science is not a storehouse of eternal truths and final facts. It might seem like it when compared to other institutions and other ways of thinking, but it is not. There are no guarantees that science will answer every question, save us from every problem, or lift us to a higher quality of existence. Science is not perfect. How could it be with imperfect humans involved in the process? Incompetent and unethical scientists are bound to be in the mix somewhere. And failure is a standard ingredient in the science recipe. Failed experiments and disproved hypotheses are the construction materials of accomplishment. Our errors remind us that the truth is still out there to be discovered.

Science may shine in the light of antibiotics, vaccines, weather satellites, CT scanning machines, and many other technologies that routinely relieve suffering and save millions of lives. But many terrors come from science as well. Scientific thinking produced the long-range artillery and aircraft used to slaughter so many civilians. Chemical and biological weapons, all products of science, can asphyxiate, burn, and infect to hellish effect. If enough thermonuclear warheads fly one day, civilization could be lost and humankind extinguished. Never forget that science is a cold, indifferent system that does not come preloaded with a moral guidance system. It is a way of discovering, learning, and applying knowledge that anyone can utilize toward any goal. If we wish, we can use science to inform our morals and elevate our behavior toward building better societies. Or, we can use it to inflict ever greater evils on ourselves. Science may exalt humanity or deliver doom. No one should worship science or revere it too much as some kind of eternal fountain of goodness. We can't count on it to bail us out of every problem. Neither can we make science the next big religion. It is no one's Oracle at Delphi or all-knowing god. It's a tool that can be used well or poorly, for good or bad. So long as humans are at the helm, the yield and quality of science will always be limited to some degree by our flaws and vulnerabilities.

None of the scientific information in this book is presented as the final word on anything. It is essential to recognize hard-won scientific knowledge as tentative. These are not permanent truths necessarily but rather the *most likely to be true at this time*. No matter how confident we feel or how mountainous the evidence may be at any given moment, it always is necessary to leave the door ajar in case we're wrong. If better evidence comes to light and upends some of the "facts" on the pages to come, then readers should adjust their thinking accordingly. I certainly will adjust mine. A competent critical thinker knows better than to remain loyal to an idea, conclusion, or theory only because it made sense in the past. Good thinking is about *now*. Does it hold up to rational skepticism and against the best evidence today? If not, it's time for a cognitive course correction—refine, revise, or reject, as needed. Be loyal only to what the evidence points to as most likely to be real and true.

Patience is an often-neglected or overlooked feature of science and good thinking in general. As important as it is to avoid blind loyalty to a conclusion, it's also a mistake to give up on one too easy. An exciting new study or discovery may seem revolutionary now but will this alleged game changer hold up? Can it survive the gauntlet of analysis and review from other scientists? It is usually wise to wait and see. Beware of exaggerations in media reporting. A single headline is not enough to justify letting go of a well-established theory or body of work that was built up over many years. Science works, but it doesn't always work as fast as we would like.

Unanswered questions far outnumber confident answers in science. Lingering mysteries are not hidden away in dark closets and ignored. Good scientists pursue them. Questions are a kind of fuel for discovery; the more the better. Lawrence Krauss, a popular theoretical physicist and cosmologist, believes we should be emphasizing the teaching of questions even more than answers.[23] Scientists are nowhere near "figuring it all out" and admit as much. Humility is a necessary component of good thinking and good science. The more you learn, the more you realize how little you know. The scientific community's openly admitted inability to explain everything is not a sign of weakness. To the contrary, it shows confidence in the system and it's one key to the great productivity of the scientific process. Science is not built on made-up facts and wished-for conclusions. Glaring gaps in our knowledge exist because good scientists don't fill them in with made-up answers. Don't allow dishonest or sincerely mistaken people to convince you that a mystery now will remain a mystery forever. If you hear, for example, someone claim that science can't fully explain X and therefore Y must be true, you should know that you are listening to a person who doesn't understand science.

Some people may wonder if "knowing things" even matters anymore. After all, it takes mere seconds for anyone with a smartphone to come up with the age of the Earth, a description of viruses, or a summary of the Big Bang Theory. Making use of the search engine Google, however, is not the same as *knowing, feeling, remembering and applying* important knowledge. The fundamental aspects of your existence deserve more than a quick Web search and several seconds of skim-reading. It is crucial to carry around within your mind, at all times, the key facts and concepts about humankind, life, and the universe. Only then is this information likely to be recalled and used when needed most, to provide a crucial frame of reference or spark of inspiration as we move through our daily lives. It doesn't matter how smart our phones are, deep knowledge can't be put on speed dial.

Never feel compelled to pretend to know something about every-thing. Ignorance is not a bad word. There is no shame in it, so long as we don't surrender and accept it as our permanent state. If not knowing something is where you happen to land in the moment, then admit it and continue to work on the problem. Confronting our ignorance—rather than denying it or camouflaging it with a quick Web search—can motivate us to seek out missing knowledge, place it in context and make some use of it. It's okay to be unsure, wait for more information, or simply accept that we have to live with an incomplete understanding of something for the time being. These may not be ideal or comfortable options, but at least they're honest.

A NEW AGE OF ENLIGHTENMENT FOR ALL

It is a mistake for an individual and any society to ricochet from task to task, from one distraction to the next, in relative blindness, never slowing down to confront the most important questions of all. The all-too common ignorance of self and one's surroundings is likely a key reason we are consistently shortsighted, delusional and destructive as a species. How many wars over the last few thousand years have been fought for senseless or trivial reasons? How many societies were driven into the ground by irrational thinking? Today we aim nuclear weapons at ourselves as if the imminent threat of global suicide is a sane state to live in. We ravage and loot Earth's biodiversity—billions of years in the making—for a brief moment of profit and pleasure. We poison and deplete Earth's massive bor-derless ocean, set fire to rainforests, and fill the sky with carbon, in large part because most people fail to think beyond the nearest visible horizon. We are very good at planning for the short-term future, but when it comes to the long-term future, however, we're not so clever. Why? In attempting to answer the question *Who are we?* we can learn that humans are the only lifeform on Earth capable of conscious, long-term future planning. We are very good at transmitting knowledge and very capable of dramatic, rapid cul-tural changes. This flexibility is of profound importance, one of the key abilities that enabled us to achieve so much. But what might we have achieved, what could we achieve, if we operate with more awareness of reality, more rational thought, and with science as a guiding force?

Overall, we behave precisely as one might expect of intelligent creatures who don't know who they are, how old they are, where they come from, how they fit in with everything else and why they think and believe as they do. Much of our behavior betrays inattention, confusion, and ignorance of basic knowledge. The prominent Harvard naturalist Edward O. Wilson describes us as something like dangerous sleepwalkers: "Humanity today is like a waking dreamer, caught between the fantasies of sleep and the chaos of the real world. The mind seeks but cannot find the precise place and hour. We have created a Star Wars civilization, with Stone Age emotions, medieval institutions, and god-like technology. We thrash about. We are terribly confused by the mere fact of our existence, and a danger to ourselves and to the rest of life."[24] The way forward should be clear enough: Learn as much of the primary and essential information about ourselves and all existence as we can manage so that it may inform and help steer our course. If we desire best-possible outcomes, for ourselves and our society, then we must first match our goals and behaviors to the best information, our best explanations of reality. And the best source for that by far is science, of course.

We will not find balance and sustainability with the natural world when most people never learn that humankind directly depends on countless other lifeforms for comfort, prosperity, and survival. We will never do our best thinking as a species so long as most of us know little about thinking. The person who has not explored the structure and function of a human brain will never discover his own mind. We will never be as large and glorious a species as we might until most of us open our eyes to the beauty and terror of the universe that is around us and within us. Ignorance of the principal questions condemns us to stumble about in a thick fog of lies, irrational beliefs, and emotional hubris. We are negligent, destructive, and foolish so often, not because we are innately evil or incurably stupid, but because most of us have not yet asked and attempted to answer: *who, what, why, when, where,* and *how.*

Chapter 2

WHEN DID EVERYTHING BEGIN?

*W**hen? When did it all happen? When did the universe, solar system, Earth and Moon begin?* These are such basic and important questions that, upon reflection, it seems profoundly odd that most people alive today probably can't answer any of them with at least a reasonable estimate. The problem with not knowing general timelines and approximate ages associated with the most momentous events ever is that it leaves people virtually incapable of forming a realistic perspective on the present. No one can ever really understand their home planet Earth, how it formed, why it functions as it does today, and how it came to be a habitat for so much life without first learning that it is billions of years old. Knowing the *when* helps us to think about the universe, Earth, and ourselves in more meaningful and useful ways. Looking up on a clear night and contemplating the universe could confuse and even intimidate many people. However, things can change with some knowledge about the universe's origin, its age, and the connections all atoms within it share, including those that construct every human. Such a moment might then just as easily ignite feelings of empowerment and inspire an injection of newfound value to the relatively brief life of a person who gets to be a thinking participant in something so vast and powerful.

This chapter is not a call to memorize specific dates. Most of the origins included here are dated by broad time ranges and estimates that could be off by many millions of years. Placing these events in context and making broader connections are the primary goals. What we want is the best evidence-based perspective possible. In the study of history, for example, specific dates matter, of course, but knowing the precise hour and day of a treaty signing or the moment the first shot was fired in some war is less important than other considerations. The priority is learning what happened, where it is falls on a timeline relative to other historical episodes, and how the event may have impacted societies and people at the time and later. Historical knowledge is most useful when connected to larger issues, events, and trends.

History is the great collection of human stories, the personal diary of modern *Homo sapiens*. Therefore, those who ignore or reject historical knowledge turn away from themselves. The past makes us, so ignorance of origins is ignorance of self. This truth does not end at the beginning of civilization, of course. Science keeps going. Archaeologists, paleoanthropologists, paleontologists, geologists, astrophysicists, and cosmologists have revealed and continue to reveal a remarkably deep past that is relevant and exciting. Humankind was not only forged and fashioned by recent history and prehistory, but also by events that occurred billions of years ago.

The formation of the universe, solar system, Earth, and Moon may have happened long before humans were around, but we exist because those things happened precisely when and how they came about. We owe everything to a moment of expansion that took us from nothing to everything, to tiny gravitational dances between clouds of rock and gas, and colossal collisions that shattered entire worlds. Every person is intimately connected with the deep past. Fortunately, with scientific methods and principles lighting the way, we all can reach back in time with our minds and begin to know when everything began.

THE UNIVERSE

Around fourteen billion years ago there was nothing. And then, according to the current most widely accepted scientific explanation of how our universe began, the endless emptiness was filled, or perhaps pushed aside, by a spectacular flash of expanding matter, anti-matter, and energy. And all this came from an incredibly small point of "infinite density". This epic enlargement of existence continues now with all of us riding through space, surfing the momentum of something that happened billions of years before the Earth or humans existed.

The word "nothing" fails to do justice to what scientists think about conditions prior to the existence of the universe. Nothing is what you might see when the lights are turned off at night in your bedroom. Nothing is what was in my bank account during my college days. This particular pre-universe kind of "nothing", however, was much more (much less?) than that. There were no stars, planets, or moons, of course. But there was *no matter of any kind*, not even a single stray atom lying about. There was no light, no energy of any kind. Finally, try to imagine this: *no time*. Before the Big Bang it seems likely that there was no past, present, or future. Scientists today still have not even figured out how to define and explain time; but whatever it is, time does not seem to have been present before the Big Bang. To sum up, then, there was less than nothing everywhere, all the time—or would have been if there had been a "where" and a "time" for all that nothingness to be. But there was one very important exception, an anomaly, perhaps, in the form of one miniscule point of . . . something. No one is sure what that it was or why it was.[25] Cosmologists and astrophysicists call this pin-point-sized something a *singularity*, which is code for: "We have no idea what it was and what was going on inside of it".

When thinking about the origin of the universe we enter a weird realm of ideas that are difficult if not impossible to comprehend. How can a dot, probably smaller than the period at the end of this sentence, contain all the ingredients necessary to make a universe? I can't even figure out how to fit all my books into a three-bedroom house with a two-car garage. Furthermore, conditions within this one tiny primordial creation capsule were bizarre, to say the least. "Infinitely hot" and "infinitely dense" is how astrophysicists describe it. I read this as more code for "unknown". But complex mystery and confusion are to be expected here. We are, after all, talking about the *ultimate beginning of everything*. You didn't think it would be simple, did you?

"It is the exception to all the rules," said Brian Koberlein, an astrophysicist and host of the fascinating podcast *One Universe at a Time*. "Every cause has an effect, but maybe the Big Bang is an effect without cause. Matter-energy can't be created or destroyed, but maybe the Big Bang arose from nothing. Events occur in space and time, but not the Big Bang. The evidence clearly points to an early hot dense state, but that raises far more questions than it answers."[26]

Do not blindly accept the Big Bang Theory just because scientists say it's true. Try to understand *why* scientists say that it's the best explanation for how everything began. What is the evidence? But expect to struggle as you try to comprehend how an entire universe burst forth from a spec. I learned the basic science behind the Big Bang long ago and it still rubs my intuition the wrong way. But that's okay because I know that science has a long and impressive record of surprising us with strange, unexpected realities that don't align well with human perceptions and expectations. The germ theory of disease, for example, describes how unseen life forms are all around and some of them can make us sick. This was not an easy claim for people to accept at first because it went against their notion of common sense. But science revealed it to be true, nonetheless. Intuition and feelings are wonderful things in their place, but they can never tell us the real story of how our universe came to be. Only by running observations, evidence, and theories through the gauntlet of science, can we do that.

Respect the difficulties of this challenge. Few things in science are as difficult as looking back nearly *14 billion years* to determine precisely what happened during those first minutes, seconds, and microseconds of the universe's existence. The Big Bang Theory is an exercise in time travel. Bewilderment and disbelief are normal reactions to this ultimate of origin stories. The brightest scientists on Earth have never fully wrapped their minds around it, so why should every layperson? Moreover, it is far from solved or settled. One could just about fill a universe with all the unanswered questions about how it was born. Anyone who struggles with the Big Bang Theory is in good company. Albert Einstein's work revolutionized cosmology to such a degree that it's almost impossible to overstate his brilliance and the significance of his contributions. But he fell short, his work incomplete. Einstein couldn't crack the origin of the universe. And, even though the Big Bang Theory is by far the most accepted evidence-based explanation, there are still constant rumblings of alternative origins among experts.

Some scientists envision a universe that has been expanding and contracting, again and again—and will continue to do so forever. Some even imagine our universe as just one of many creation events, one big bubble in an endless froth of universes. And, ponder this one for a while: If there are an infinite number of universes (the multiverse hypothesis) it probably means that there are universes somewhere that contain *anything* and *everything* you can imagine. But these possibilities, interesting as they may be, cannot match the strength of evidence behind the Big Bang Theory.

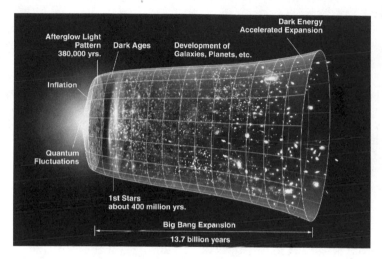

Image by NASA/WMAP

Those who may feel that the Big Bang Theory is a nothing but a collection of ridiculous ideas, magical thinking, or just plain impossible, might consider that this is no standard human origin claim. It is unique, nothing like the millions of creation myths that countless cultures have offered up over the last several thousand years. This one came from science which means that it has something very special going for it. It is not the result of inspired guesswork or creative storytelling. And the tale isn't conjured up by scientists out of thin air so that they might win grant money or feel better about the universe they find themselves trapped inside of. It may turn out to be wrong but the reason to accept it now are correct. It is based on evidence. The claims put forth by the Big Bang Theory do not rest on the weighted words of special people. The key figures who contributed to the theory—Alexander Friedmann, Robert Dickie, James Peebles, Albert Einstein, Georges LeMaitre, Edwin Hubble, Alan Guth, and others—were not gods or prophets. They were not visited by angles or aliens and handed the key equations. This was a case of mortals putting in the necessary hard work, struggling with complex numbers, thinking up experiments, and deciphering observations. As a result, there are very good reasons why the Big Bang makes sense—even if it sometimes feels like fantastical nonsense.

NOT THE BEGINNING

The Big Bang does not mark the actual beginning of the Universe. Most people seem to miss this, but the theory explains what happened from a small fraction of a second *after* the expansion began and onward. "[T]he big bang leaves out the bang," writes Columbia University professor of physics and mathematics Brian Green in his book, *The Fabric of the Cosmos*. "It tells us nothing about what banged, why it banged, how it banged, or, frankly, whether it ever really banged at all."[27] The reason the Big Bang is central to discussions about the beginning of the universe is that it is the closest to the beginning that science currently can take us. But it's not the beginning. Cosmologists and astrophysicists freely admit this, of course, and keep working on the problems in hopes of getting closer to that ultimate moment of creation, and even maybe a bit before it as well, if that is possible.

Sean Carroll, a theoretical physicist at the California Institute of Technology, describes the Big Bang as "the end of our theoretical understanding. We have a very good idea, on the basis of observational data, what happened soon after the Bang. The microwave background radiation tells us to a very high degree of precision what the universe was doing when it was a nuclear fission reactor, just a few minutes afterward. But the Bang itself is a mystery."[28]

There is a similar misconception about biological evolution that is common. The Theory of Evolution explains how life changes over time—after life existed. It does not specifically address the beginning of life. For this reason, almost all those evolution versus supernatural creation debates are askew from the start. Such debates should be described as *natural origin* versus *supernatural origin* or *life changes* versus *life is static*.

The Big Bang is often dismissed as *"just* a theory" by people who don't understand or like it. The theory of Evolution suffers this same problem. Contrary to a common belief, however, a "theory" in science is something very special; it's respectable, potent, and valuable. The Big Bang is a *scientific* theory and that is significant. It means something. The word theory may be used in popular language to refer to a mere guess or some flimsy, half-baked idea, but scientific theories are nothing of the sort. They are a destination, an achievement, reached only after a lot of rigorous work and testing. Theories prove their worth not only by an accumulation of evidence, and stand up against attempts at disproving them, but also by being able to make accurate predictions. This is the sweet spot of science, one of the key reasons it is so productive at enabling us to do real things in the real world. Prediction from theory is how science beats magic and wishful thinking when it comes to our being able to fly, communicate with people on the other side of the planet, or do millions of other amazing things.

Want to test the theory of Evolution? Predict that fossils of older, extinct plants and animals will be found in older rock strata and fossils of younger life forms will turn up in younger rock strata. Then go out and find some fossils to see if the prediction holds true. Want to test the Big Bang theory? Predict, in accordance with the Big Bang model, that there are twelve hydrogen nuclei for every helium nucleus in the universe. Add to that a prediction of forty thousand hydrogen nuclei for every deuterium nucleus, then check to see if it's true. Scientists did this and it turned out to be exactly right.[29] The Big Bang theory, for all its confounding weirdness and impenetrable mysteries, has enabled scientists to make many accurate predictions and has survived decades of challenges from experimentation and observation. It remains neither perfect nor complete, but it's still standing.

The best observable evidence for the Big Bang is the "noise" left behind in the form of radiation. This evidence was picked up in 1965 by accident. Arno Penzias and Robert Wilson were conducting experiments, unrelated to the Big Bang, using a radio telescope at the Bell Lab in New Jersey and detected radiation in the form of cosmic microwave background (CMB) coming from all directions and at all times. CMB is exactly what should be there. It is the remnant heat of a super-hot newborn universe, there just as predicted by the Big Bang Theory in 1948.[30] Penzias and Wilson shared the 1978 Nobel Prize in physics for their discovery.

DARK UNIVERSE

Everyone knows that an overwhelming majority of stars, planets, moons, asteroids, comets, and black holes remain unseen and unexplored. But our current ignorance runs far deeper because all those things account for *less than five percent* of the contents of our universe. The rest is a total mystery. Data from the Hubble Space Telescope in 1998 revealed that something very strange was out there and everywhere, to a very large degree. Gravity is not slowing the universe's speed of expansion as expected. Instead everything seems to be speeding up.[31]

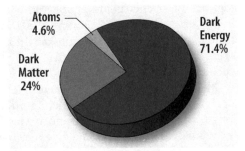

Atoms 4.6%
Dark Energy 71.4%
Dark Matter 24%

NASA/WMAP Science Team

There is just no good available explanation at the moment. Scientists have named it, whatever "it" may turn out to be, *dark energy*. We can be sure that dark energy is significant because it makes up about 70% of the universe. This is calculated by looking at its influence on the standard matter in the universe (stars, planets, etc.). But if 4% or 5% of the universe is regular matter and about 70% dark energy, then what is the remaining 25% percent or so? Things get even stranger. Based on theoretical models and observations, there is more matter than there should be, or at least more than we can detect. This is another profound mystery. Approximately a quarter of the universe's matter seems to be some substance or substances other than stars, planets, clouds, black holes, people, rocks, or anything else we are familiar with.[32] But, as with regular matter and dark energy, *something* is there.

John Bochanski, a physicist and popular blogger for *Sky and Telescope* magazine, describes the Big Bang as a challenging concept for anyone to think about, but worth the effort. "The Big Bang is one of the most difficult events to wrap your mind around," he said. "How does the vastness of our Universe start with a single point? The early Universe is so wildly different from our current experiences that it is easy to see why some might find it bizarre."

Bochanski stresses that the Big Bang is based on evidence:

> Decades of observations and theoretical efforts have led to a fundamental understanding of our expanding Universe during its infancy," he explained. "Our understanding is directly related to observations of the Cosmic Microwave Background, seen in every direction with essentially the same brightness. It is the echo of the Big Bang and observations of this echo have led to our best understanding of the structure of our Universe. The structure predicted by ripples in the Microwave background match the distribution of galaxy clusters throughout the Universe, measured by very different experiments and surveys. These pictures of our Universe lead to a fundamental understanding of our Universe and its formation and evolution. The human mind is good at relating observations to itself, and to perceiving short, that is, hour, day or year timescales. It struggles, or at least mine does, at relating what I observe today to events that have happened centuries or millennia ago. I know that small rivers lead to the grand canyons, that a single cell can lead to a person, and that dinosaurs once roamed my hometown. But I cannot observe any of these processes directly. Instead, my knowledge is rooted in centuries of observations and scientific analysis, developing a rich portrait of our Universe and our place in it.

A NASA research satellite, the Wilkinson Microwave Anisotropy Probe (WMAP), delivered strong data in 2012 that further supported the Big Bang model. The probe mapped small temperature fluctuations across the universe, the "afterglow", left behind by the Big Bang. Based on the findings, the WMAP's science team dated the universe to 13.77 billion years ago.[33] The European Space Agency also measured this relic radiation with its Planck spacecraft, determining an age of 13.799 billion years for the universe.[34]

It is amazing how fast our view of the universe has improved in little more than a hundred years. As recent as the earliest years of the twentieth century, most scientists thought that there was only one galaxy, the Milky Way that we inhabit, and nothing but empty infinite space beyond. Einstein's famous theory of General Relativity, published in 1916, predicted the Big Bang theory and a dynamic, expanding universe, in total conflict with the accepted view at the time of a stagnant universe. Einstein's numbers were too far ahead of the best observations to be made sense of. Attempting to resolve the problem, Einstein added the "Cosmological Constant" to his theory to allow for a stable and somewhat stagnant universe. But in the mid-1920s, astronomer Edwin Hubble observed other galaxies that were moving away from our galaxy at speeds proportional to their distance from us. This matched predictions made by the original theory of General Relativity and helped set the stage for later broad acceptance of the Big Bang model.[35]

Einstein regretted not having more faith in his original conclusions and called his late insertion of the "Cosmological Constant" the "biggest blunder" of his career.[36] Blunder or not, thanks to the work of Einstein, Hubble and many others, we now know that there are *hundreds of billions* of galaxies in the universe, each containing many billions of stars—and they are on the move. These movements, measured and tracked back through time, are powerful evidence of the Big Bang theory and a universe that is 13.8 billion years old. An interesting postscript to the Cosmological Constant is that it remains relevant and may prove useful in helping to make sense of the effects of the mysterious Dark Energy that pervades the universe.[37]

Why does any of this matter? What does knowing the age of the universe mean? How can something so long ago possibly have relevance now or bring any more meaning to the life of an individual human in the twenty-first century? It matters because the beginning of the universe was the beginning of you. Your body is mostly hydrogen atoms, created in the early moments of the Big Bang. The other atoms that make you came from stars that only existed because the Big Bang happened and played out precisely as it did. Awareness of the Big Bang presents the profound realization that you are far older than you may ever have imagined. We all have two birthdays, one could say. The most recent, of course, happened when our mothers gave birth to us. The first, however, happened almost fourteen billion years ago in that spectacular flash of expansion. It was the moment when everything began. And we, most definitely, are a part of everything.

THE SUN

The first and most prominent resident of our solar system is the Sun. All humans live within its gravitational grip and depend on it for survival every moment of their lives. It seems reasonable, therefore, that everyone ought to know at least a little something about it. This star, a middle-aged G-type main-sequence star, formed around 4.5 to 4.6 billion years ago in the Milky Way galaxy.[38] In a crucial moment, certainly for future *Homo sapiens* anyway, a cloud of gas and dust, almost all of it hydrogen and helium atoms, became sufficiently dense that gravity caused it to begin collapsing in on itself. During this process, the material spun in a huge disc, capturing and sucking in more and more materials over time. As this went on for millions of years, the material accumulated and thickened around a central point, heated up, and eventually became so hot that nuclear fusion occurred. And a star was born. This may have only been one among billions of stars in the Milky Way galaxy, which was just one galaxy among billions of others, but it was special. This was the star that would deliver the energy necessary for life to rise and flourish a billion years or so later on a little world that had yet to form from leftover debris.

Nuclear fusion happens when a single heavier nucleus forms from two lighter nuclei being forced together. Excess energy is released during this process. In the case of stars, hydrogen and helium atoms continually slam into one another and, all together, release astonishing amounts of energy. According to NASA calculations, for example, the Sun coverts four million tons of matter into "pure energy" *every second*.[39] So think of the Sun as a colossal fusion bomb with important consequences for the universe. Remember that hydrogen atoms were made during the Big Bang, but all others come from fusion inside of stars. These building blocks—necessary to make a planet, moon, trees, or you—came from the furious furnaces of countless stars throughout the universe.

By human standards, the Sun is extremely old, large, and hot. This star accounts for about 99.9 percent of all the mass in the solar system.[40] Yes, strange as it may seem, all the planets, every moon, all the asteroids, rocks and dust make up only about *a tenth of one percent* of the solar system's total mass. Given its role as the crucial source of energy for Earth life, and in determining the location and movements of everything else in the solar system, it should not surprise us that more than a few cultures have regarded the Sun as a supernatural god worthy of worship.[41]

MORE THAN FIRE

- The Sun is about **4.6 billion years old** and 93 million miles (150 million kilometers) away from the Earth on average.
- It takes about eight minutes for **light from the Sun** to reach the Earth.
- The Sun is big enough to contain about **1.3 million** Earth-sized planets.
- The Sun's nuclear fusion reactions produce approximately **386** *billion billion* megawatts per second.
- Ninety-one percent of the Sun's atoms are **hydrogen**. Helium is the next most common atom at nearly 9%. Oxygen, nitrogen, carbon, and some others are present in very small numbers.
- Hydrogen makes up a smaller ratio of the Sun's **total mass**, 71%, because it's so light. Helium is at 8.7% and oxygen a little less than 1%.
- The Core, Radiative Zone and Convection Zone make up the Sun's **inner layers**. The **outer layers** are: Photosphere, Chromosphere, Transition Region, and the Corona.
- The energy created inside the Sun takes about **a million years** to reach the Earth in the form of warmth and light.
- Pressure at the Sun's core is **340 billion times greater** than the Earth's air pressure at sea level.
- It takes **several hundred thousand years** for radiation to go from the Sun's core to its upper layer.
- The Sun's core has a temperature of about **15 million degrees Celsius** (27 million degrees Fahrenheit). It makes up only 2% of the sun's volume but the core accounts for nearly half of the Sun's mass.
- The Sun's visible areas are its **coolest** areas with temperatures of approximately 6,000 degree Celsius (10,832 degrees Fahrenheit).
- The Sun is one of approximately **70 billion trillion** (7×10^{22}) stars.[42]
- The Sun will **not die as a supernova** in a massive blast. Only about one percent of stars die this way. The Sun will go out as a white dwarf, gradually cooling down over billions of years.

Source: NASA[43]

We owe a lot to the Sun—everything, one might argue. Its gravity currently holds the Earth in a steady orbital track that keeps it at just the right distance, just warm enough and with just enough light, for life as we know it to exist. Don't make too much of this favorable location, however. If the Earth were closer to or farther away from the Sun, it simply would mean that different kinds of life would have evolved on it or no life at all would have evolved on it. The Earth's unique position in the solar system does not prove the existence of an intelligent cosmic designer, as many have claimed, any more than the unique positions of Venus, Neptune, and any other planets presumed to be inhospitable to life disprove the claim. Additionally, the impressive flourishing of so much biodiversity does not mean the Earth is a perfect paradise. That should be clear with devastating climate swings and asteroid strikes in the past, and an overall species failure rate estimated at 99 percent.[44] Life may be easier here than on many other planets, but that doesn't mean it's easy.

THE EARTH

If you are outside right now, or the next time you are, look around. I mean *really look around*. Notice the trees, hills, mountains, buildings, or whatever is near you, glance up at the sky, look down at your feet on the ground, and then think about this: You and everything around you are riding atop the ultra-thin crust of a spherical, mostly molten rock, natural, ancient, magnetically shrouded, atmosphere-shielded spaceship—and it is ripping through space at amazing velocities. Most people probably never think much about this but it's stunning to stop and ponder how fast and in how many directions we are tearing through space every moment. I like to imagine observing the Earth from a stable position in space as it flashes by in a blue-green blur.

According to NASA, the Earth spins at a speed of 1,037 mph (1,670 kph) at the equator.[45] This spin rate is so fast that centrifugal forces cause the planet to bulge outward.[46] But that's far from the whole story. The Earth also orbits the Sun (one revolution per year, of course) at a speed of about 67,000 mph. Meanwhile, the entire solar system, which includes the Earth, is ripping through space around the center of the Milky Way galaxy at a speed of about half-a-million mph.[47] And, finally, our galaxy is moving through the universe at an extremely high rate of speed as well. We aren't all space sick or pinned to the ground by G forces because we and everything else on the Earth is moving at the same speed. It's like being in a fast-moving car or airplane. Once you get going, on the highway or at cruising altitude, you don't sense steady forward movement anymore. You only feel significant accelerations, decelerations, sharp turns, or bumps. If the Earth were to suddenly speed up, slow down, or hit a significant bump, everyone would notice to say the least.

HOME

- The Earth is the **largest of the four terrestrial planets** (rocky worlds) of the solar system and fifth largest of all eight planets.
- **The inner core**—a solid sphere about the size of the Moon and made of iron and nickel. It is the deepest part of the planet and extremely hot (9,800 degrees Fahrenheit/5,400 degrees Celsius).
- **The outer core** is fluid, 1,400 miles or 2,300 kilometers thick, and envelopes the inner core. It's made of molten iron and nickel. Together, the inner and outer account for 15% of the planet's total volume.
- **The mantle** is molten rock some 1,800 miles or 2,900 kilometers thick. It makes up 84% of the Earth's volume.
- **The crust** is the upper part of the Earth. On land, it's about 19 miles or 39 kilometers thick on average. On the ocean floor, however, it's much thinner, about three miles or five kilometers thick. The crust makes up **a mere one percent of the planet's volume**. The shell of a chicken egg, for comparison, is relatively much thicker than the crust.
- The Earth completes one rotation every **23.9 hours** and takes **365.25 days** to make one trip around the Sun. That extra quarter-of-a-day per year is the reason a day is added every four years (leap year).
- Earth's atmosphere is **78% nitrogen, 21% oxygen**, and 1% other elements. It is the ideal balance of gases for humans to

breath because it is the atmosphere that humans evolved in.
- The Earth's thin atmosphere (most of it within 10 miles/16k of the ground) **protects life** from many incoming objects from space. Most of them break up in the atmosphere due to friction/heat.
- An **ocean** covers about 70% of the Earth's surface.
- The Earth's **radius** at the equator is 3,958.8 miles or 6,371.00 km. The **circumference** at the equator is 24,873.6 miles or 40,030.2 km.
- **Surface temperatures** on Earth can dip as low as -126 degrees Fahrenheit and as high as 136 degrees Fahrenheit. (Celsius: -88 minimum and 58 degrees maximum)
- The Earth is the only planet in the solar system not named after a god. "Earth" is a **German word for "ground" or "soil"** and has been in use for at least a thousand years.
- The Earth is the only planet in the universe **known to have life.**

Source: NASA, USGS[48]

The Earth is old. *Very old*. It formed about 4.54 billion years ago.[49] The late astronomer Carl Sagan's classic "cosmic calendar" from the 1970s still works well for gaining perspective on just how relatively long ago or recent important things happened.[50] For example, if all time were compressed into a single Earth year and the Big Bang (13.8 billion years ago) happened on the first day of the year, January 1, then our Milky Way galaxy wouldn't form until May, five months later. The solar system and the Earth come four months after that on September 1. The first Earth life appears around mid-September. Animals are on the rise in early December and December 26 marks the arrival of dinosaurs. Mammals show up two days later. All non-avian dinosaurs are extinct by December 30 after an asteroid impact. At 9:30pm on December 31, bipedal apes are on the move, walking in Africa and then, only a couple of hours later at 11:52pm, anatomically modern humans are living in Africa. Agriculture begins catching on with just 23 seconds left in the year. Finally, all modern history occurs in the last second of the cosmic year.

Understanding the immense age of the universe and the Earth is crucial knowledge because it touches and leads to so much more.

Knowing that the Earth is *billions of years* old is a vital prerequisite to having a fair chance of understanding a long list of important topics. No one can have a sensible perspective on Earth life, for example, if he or she does not first know that the Earth is extremely old. It is impossible to gain a realistic view, or make the most productive use, of geology, obviously, but also archaeology, astronomy, astrophysics, botany, cosmology, entomology, genetics, herpetology, marine science, marine biology, paleoanthropology, paleontology, primatology, and zoology. All of these fields of science, and more, have proved themselves to be consistently productive at revealing new aspects of reality. And all of them begin with a very old Earth as a frame of reference. If the Earth were young, most of current science would be wrong and simply could not work. Plate tectonics, evolution, the formation of mountain ranges and the ocean, prehistoric human migrations, and so on could not have happened as understood on a freshly minted Earth. In addition to the overwhelming circumstantial evidence, however, there is something better: direct evidence that shows the Earth to be about four and half billion years old, with 99% confidence.[51] Prominent geologist G. Brent Dalrymple, now retired, is a winner of the National Medal of Science. He also is the author of an important book on this topic, *The Age of the Earth*, published in 1994.[52] In a 2007 essay, Dalrymple declared: "There is universal agreement amongst knowledgeable scientists that the earth, the solar system, the galaxies, and the universe are billions of years old. . . . This evidence is so conclusive and scientists are so confident about the antiquity of the earth . . . that it is not a subject of debate within the scientific community. Furthermore, there has been no disagreement about this subject for more than half a century . . . "[53]

Scientists use radiometric dating to determine the extreme ages of very old rocks and minerals. This allows them to deduce a minimal age of the Earth. Developed in the early twentieth century, radiometric dating, has proven to be reliable and accurate. It works by measuring the radioactive half-life of isotopes in the rock or mineral. This refers to the original or "parent" radioactive isotope's rate of decay into the new or "daughter" isotope.[54] Using radiometric dating techniques rocks older than 3.5 billion years have been found on every continent. The oldest known Earth rock fragments found and dated so far are Canada's Acasta Gneiss at a little over four billion years old.[55]

OLD EARTH. Canada's Acasta Gneiss rocks are more than four billion years old, according to reliable radiometric dating. *Photo by Emmanuel Douzery*

It is naturally occurring lead isotopes that provided a key source for that Earth birthday date of 4.54 billion years. According to the United States Geological Survey (USGS), "the ratio of lead-207 to lead-206 changes over time owing to the decay of radioactive uranium-235 and uranium-238, respectively."[56] The USGS continues:

Scientists have used this approach to determine the time required for the isotopes in the Earth's oldest lead ores, of which there are only a few, to evolve from its primordial composition, as measured in uranium-free phases of iron meteorites, to its compositions at the time these lead ores separated from their mantle reservoirs. These calculations result in an age for the Earth and meteorites, and hence the Solar System, of 4.54 billion years with an uncertainty of less than 1 percent. To be precise, this age represents the last time that lead isotopes were homogeneous throughout the inner Solar System and the time that lead and uranium was incorporated into the solid bodies of the Solar System.

Many millions of people worldwide currently imagine that the universe, Earth, and/or the human species have very recent origins, measured in mere thousands of years rather than billions.[57] This disconnect from an overwhelming convergence of scientific evidence can be traced to some versions of some religions continuing to encourage or demand that their adherents accept an absurdly young Earth that is only 10,000 years or so old. But no rational case can be made for believing this. It also seems unnecessary as many apparently sincere and practicing Christians, for example, seem to accept the real age of the Earth without any negative consequences.[58] This includes both Protestants and Catholics. Pope Francis stated in 2014 that the Big Bang, the evolution of life, and an old Earth do not contradict his belief in God.[59]

The case for a 4.54 billion year old Earth is solid, having been made by radiometric tests dating some rocks to be older than four billion years and zircon crystals to be 4.4 billion years old. This evidence adds up to an ancient planet, one that has been around for at least 4.4 billion years old.[60] The older and precise age is the result of both evaluating Earth materials as well as evidence from meteorites and the solar system at large that the Earth formed in conjunction with. It can be said now, beyond any reasonable doubt, that the third planet from the Sun is four-and-a-half billion years old.

CORE CURRICULUM

Few people probably give it much thought, but humankind exists in large part courtesy of amazing, *life-enabling* activities occurring deep within the Earth. Learning about these important subterranean processes can help us to further understand and appreciate the vital connections we share not only with other Earth life but with the planet itself.

The crust, where everything lives, makes up just 1% of the Earth's total volume. Though incredibly thin relative to the overall Earth, we have never been able to penetrate it to directly explore the mantle beneath. It's just too hot and difficult to reach. Some might say impossible to reach. It is remarkable and probably surprising to most that we have not directly studied 99% of our own planet. NASA's *Voyager 1* probe is more than 13 billion miles from Earth, travelling at about 38, 000 mph, and still exploring.[61] Meanwhile, we still have not yet been able to physically probe the mantle that lies a mere five miles beneath the ocean floor and twenty miles or so deep on land. The inner core may as well be on the other side of the galaxy. Geologists are no less motivated in their work than space explorers, of course. It's just that it's a lot easier to move through space out there than it is solid rock and molten rock down here. Nevertheless, by doing science—analyzing the chemical composition of rocks, studying how energy waves from earthquakes behave as they move through the Earth—a lot has been learned about what is going down below.[62]

WHY ARE THERE SEASONS?

Temperatures in various parts of the Earth don't stay the same year-round because *the Earth rotates with a 23.4-degree tilt in relation to its orbital plane around the Sun.* This causes different areas to get different amounts of sunlight and heat at different times. For part of the year, the Earth's northern hemisphere is tilted toward the Sun and gets more heat (summer). During this same time the southern hemisphere is tilted away from the Sun, making it colder (winter). This then reverses for the other half of the year.

Probably the most relevant thing to know about the inner Earth is that we all would be dead without it. Or, more precisely, humankind would never have existed in the first place without it. The ingredients and movements of the constantly swirling outer core generate a magnetic field that deserves our deepest gratitude.[63] This energy field expands through the Earth and beyond to shield the entire planet. It protects humans, and most other Earth life, from deadly cosmic and solar radiation. It also helps gravity hold the atmosphere in place and keep it from leaking away into space.

We may owe everything to the crust and mantle, too. Although the continents might seem fixed and permanent to us, they are far from stationary. The surface is in constant motion as giant slabs of crust float, slide, collide, and rub one another atop an ocean of melted rock. All this slow-motion (from our perspective) movement is called plate tectonics and it could be an essential key to life on Earth, possibly on other planets, too. Moving plates stir up and replenish the nutrients that early microbial life would have needed to survive on a young Earth.[64] The process recycles carbon also, important for stabilizing temperature. And, finally, plate tectonics contributes to the generation of the Earth's magnetic shield.[65]

THE MOON

The Moon is another significant component of immediate human reality that deserves more attention than it gets. After all, this nearby world provides a few important services for humans and other Earth life. And, if that's not interesting enough, its scientific origin story is nothing short of spectacular. The Moon has benefited us by helping to stabilize the Earth on its 23.4 degree tilted axis. The reason this matters is that it helps make Earth's climate more stable than it otherwise would be. The Moon is also the primary cause of regular tides. Its gravitational force pulls the Earth's water back and forth. This matters today because intertidal ecosystems are a significant part of our global biodiversity. For many species of plants and animals, the tides are life.[66] But tides could be even more important. They might be a key reason there is any life on Earth at all.

Molecular biologist Richard Lathe was instrumental in the development of the vaccine that eradicated rabies in several European countries. He also has contributed to the investigation of life's origins. He believes the Moon was a catalyst for life's start. Four billion years ago the young Moon was closer to Earth and its gravitational pull, combined with the young Earth's faster rotation, caused more extreme and more frequent tides. This, Lathe states, "could have provided a driving force for cyclic replication of early biomolecules."[67] In addition to simply churning up the primordial soup, Lathe thinks that the more active ocean, with tides occurring every two to six hours, caused repeat high and low saline environments. This could have led to new molecular combinations—one or more of which may have kickstarted life. Lathe told me that it's very important to recognize that there likely was more than one *origin of life* on Earth, whether it occurred in deep-sea vents or as a result of tidal action. "The emergence of replicating nucleic acids is one, and of lipid-coated micelles another, and of cellular metabolism," he said. "It must have taken a while for all three, or more, to have got together."[68]

The Moon has been with the Earth from almost the beginning. While the Earth's age dates to 4.54 billion years ago, and the Moon was relatively close behind at 4.41 billion years old.[69] Experts once thought that the Moon's existence was the result of the Earth capturing passing materials with its gravity and locking the debris into permanent orbit where it then compacted to form the sphere we know today. Others thought the Earth and Moon had formed independently at the very start of the solar system, bound together by gravity from the start. A third idea proposed that the Moon came from a colossal chunk of the Earth that had broken off early in Earth's formation. The big problem with these possibilities is that the composition of the Earth and Moon are very similar but not identical. This seems to argue both for all of the above and none of the above. If the Moon is a big piece of the Earth, then there should be no differences in the composition of their rocks. But there are. On the other hand, if the Moon formed apart from the Earth, then moonrocks should have their own unique signature. But they don't.

THE LONELY NEIGHBOR

- The Moon is approximately **4.51 billion years old** and is the Earth's only natural satellite.
- The Moon has a radius of 1,079.6 miles or 1,737.5 km. It's **less than a third of the Earth's width**.
- The Moon is the **fifth largest moon** in the solar system.
- There are more than **180 other moons** in the solar system.
- The Moon orbits the Earth at an average distance of about **239 thousand miles** or 384 thousand km. The extreme ranges from the Earth are from 356,400 km at closest, 406,700 km at farthest.
- The Moon helps **reduce the Earth's wobble** on its axis. This helps stabilize the climate.

- The Moon is the primary cause of regular **ocean tides** which some scientists think were critical to the earliest life on Earth.
- Only **one side of the Moon** is ever visible from the Earth because it spins at a rate equal to its 27-day orbit around the Earth.
- The Moon was probably created from debris ejected by Earth's collision with a Mars-sized object, named **Theia**. Moon rocks retrieved by the Apollo missions are mostly Earth-like but with traces of some other object, as would be expected by such an origin.
- The Moon has an extremely thin atmosphere, called an **exo-sphere**.
- The Moon's orbit is slowly expanding, moving it **farther away from the Earth** at a rate of about one inch per year.
- NASA made **nine voyages to the Moon** in the late 1960s and early 1970s. These included six lunar landings. In total, twelve humans have walked on the Moon.
- Although American astronauts planted flags on the Moon, the United States did not "claim it". The US agreed to **a 1967 international law** that prohibits any nation from owing stars, planets, moons, and other natural objects in space.

Source: NASA[70]

Today most scientists reject those three concepts in favor of the *giant-impact hypothesis*. It's a work in progress, for sure, but the most widely accepted explanation because it seems to best fit the evidence.[71] It describes a Mars-sized object, a protoplanet named Theia, striking the young and still-forming Earth. The impact send debris into Earth orbit that would then form the Moon. This happened when the Earth was only about 50 to 100 million years old and much more dense and softer than it is today. The impact was incredibly powerful, producing perhaps *100 million times more energy* than the asteroid impact thought to have devastated the dinosaurs billions of years later.[72]

Try to see this collision in your mind. The Earth and another planet or object, *the size of Mars*, hurtle toward one another. They are on a collision course that was locked in by an unknown series of bumps and spins, as well as gravitational tugs and nudges, most of them likely occurring millions of years before this moment. Imagine yourself hanging out in space, observing the imminent impact from a safe distance. When they finally touch, millions of tons of rock begin breaking up and flying off into space. There are no accompanying "booms" or loud explosions because sound does not travel in airless space. To see all this carnage on a grand scale occurring in total silence seems impossibly weird.

VIOLENT BEGINNINGS. Artist's conception of two worlds colliding. The Moon is thought to have been created by an impact similar to this. *NASA illustration*

For a moment the spectacle feels peaceful, beautiful. But the noise of your own frantic breathing inside your space helmet reminds you to be afraid. The two giant spheres continue their merge until they have morphed into one misshapen mass, with the Earth absorbing much of Theia's mass. Colossal chunks of rock rise from the chaos, shooting off in all directions. Jets of debris spew wildly out into space from the Earth on the opposite side of the impact point. Then you notice a barrage of boulders heading directly at you and realize that there is no such thing as a safe place during the solar system's infancy.

EVIDENCE OF IMPACT. Astronaut John W. Young, commander of the Apollo 16 lunar landing mission, collects Moon rock samples on April 23, 1972. Moon rocks and Earth rocks have nearly-but-not-quite identical chemical signatures which supports the hypothesis that a large object struck the Earth and the Moon formed from the resulting debris. *Photo by Charles M. Duke Jr./NASA*

Most experts favor the giant-impact hypothesis today because it helps make sense of the rocks. If a massive object struck the Earth long ago, as described by this idea, then the Moon-making debris ejected during the impact should be mostly of the larger Earth but with a little something from another object mixed in. This is the telling clue. It does not close the case by any means, but it seems to explain why Moon rocks brought back to Earth for analysis more than four billion years later by Apollo astronauts would be almost identical to Earth rocks in some respects yet just different enough to indicate that something other than the Earth had a hand in the Moon's origin.[73] Perhaps that something was Theia.

This chapter attempted to answer *when* everything began. While nothing is ever perfect or guaranteed forever, science has done well for us in this regard. The evidence is strong for a 13.8 billion-year-old universe. The Sun and solar system date to about 4.6 billion years. We can be very confident in a 4.54 billion age for the Earth and the Moon is looking good at 4.51 billion years old. One or all of these dates could change, of course, as more discoveries are made and our understanding of the universe improves. But none of them are likely to change by much. We can be very confident that we live on an old planet with an old neighbor, the Moon, nearby. There is no reasonable doubt that our universe is anything but immensely old. These dates are only part of the story, however. The *when* connects to the *why* and the *how*. And, scientific progress always generates more questions than answers, just as it should.

What, if anything, existed before the Big Bang? The work of physicists has shown us that "nothing" just isn't what it used to be. What, if anything, was going on within all that timeless nothing before the Big Bang? Why did the Big Bang occur at all? How *exactly* did the Earth and other planets form? Is the Moon really the result of one of the solar system's all-time great collisions or was its origin more peaceful? And, we now have dark matter and dark energy to contend with after discovering that approximately 95 percent of the universe is a big fat unknown. But there is no need to let unanswered questions frustrate us too much or mislead anyone into viewing science as inadequate or having failed. Big mysteries are the precise reason we need the tools and principles of science. Science is the only way we will find those answers. The technology used to detect and measure cosmic background radiation was not wished into existence. Hopes and fears did not gift us the age of the oldest Earth rocks and crystals. Radiometric dating did.

Where would we be now without our current accumulation of scientific discoveries and theories? It's scary to imagine how little evidence-based knowledge we would have about everything and how much confidence there would be in mistakes and lies without science. The big unsolved riddles may frustrate and irritate us but they are not all bad. They can be high-octane intellectual fuel for the individual and for all humankind. They are the positive motivation for us to keep thinking and keep seeking.

Chapter 3

WHO ARE WE?

Take a long, hard look into the metaphorical mirror. Who are you? Don't answer too fast, however. Go for substance. Cut through the surface noise. Move above and beyond name, gender, nationality, skin color, and any religious or political factions you may identify with. "Human" is the quick and obvious answer, of course. But what does that mean? What is a human? This is a crucial and necessary question. Attempting to answer it, with both the powers and constraints of science, can help us become more self-aware, grounded, and sensible. So, then, *who are we?*

We are animals, mammals, primates, people, *Homo sapiens*. But those are just labels. What is the meaning that lurks behind them? If a team of visiting extraterrestrial zoologists were to ask you to define or explain us, you certainly wouldn't satisfy their curiosity by simply answering "human" and walking away. Much more information is required to make some sense of us. What are we made of? Where did we come from? How long have we been here? What are our needs? What is our relationship with other Earth life? What do we do with our time and energy? What do we think about? What does it mean to be a human now compared with being a human two million years ago? Who and what were our ancestors? What possible futures are ahead of us? *Who are we?*

Humans are one version of life on one planet. We are a single lifeform among an estimated trillion or more species.[74] But what a special one we are. For better and for worse, we are unique. A human is a big-brained, bipedal, highly sexual, omnivorous, skin-covered, bone-supported, microbe-hosting, social, daydreaming, verbose and gossipy, information-sharing, family-loving, playful, creative, art-loving, religion-prone, materialistic, planning, destructive, world-altering, spacefaring, flexible, culturally amplified complex cellular organism. The world and, maybe, the universe have never seen anything like us. Our thoughts and dreams can be extraordinary. Right now, for example, some members of the human species are working hard trying to figure out how to defeat death, teleport matter, build computers with god-like abilities, and how best to colonize faraway worlds. The species *Homo sapiens* has many faults and failures that cannot be fairly ignored. We have a capacity for great love and hate, great compassion and great cruelty. We both build and destroy. Still, it is undeniable that humankind is something special, a creative, courageous, and capable lifeform with a knack for solving difficult challenges, perhaps possessed even by a bit of a compulsion to pursue improbable and impossible goals.

Apart from being matter and alive, the most basic qualifying description of a human is Eukaryote (pronounced "yoo-kar-ee-oht"). This classification means people are among the relatively complex lifeforms, more closely related to mushrooms, ants, and redwood trees than to the simpler Prokaryotes which include the old, prolific, and tough bacterial species. Humans are vertebrates, animals with a backbone. We also are mammals, warm-blooded, hairy milk-producers. Unlike cold-blooded reptiles, humans are endothermic animals. We generate and regulate our own body heat. This has helped mammals to live in diverse environments, and was a key advantage during our precarious prehistoric years when various species of humans spread throughout much of the world. It is easy to see that modern people are related to diverse lifeforms all around. Many physical similarities to other primates (such as chimpanzees and gorillas) are obvious and our genetic closeness to almost all life on Earth is even more remarkable. But let's not get ahead of ourselves. To know who we are requires understanding a few things about the fundamental substances and structures of a human body.

WHAT ARE HUMANS MADE OF?

Humans are matter. We exist as physical, tangible entities, something more than an idea, for example. A human might be thought of, minimally, as a collection of chemical reactions that fall within a particular range. Being matter means we are atoms, of course, because all matter is composed of these primary building blocks. Contrary to what our eyes and common sense may tell us, a human is mostly empty space because atoms are mostly empty space. No less odd than that is the fact that our bodies are always changing. The atoms that constitute "you" never cease coming and going throughout your lifetime. You are not the same collection of atoms, molecules, and cells you were at birth. You are not the same configuration you were yesterday, or even a second ago. Therefore, it makes sense to think of ourselves as *persistent patterns*. Humans grow and mature, interact with their environments, consume and excrete things, injure themselves, suffer infections, age, and die. All this activity keeps our patterns in flux; change is the permanent feature of our existence. Every human is a unique but fluid arrangement of atoms.

How many atoms does it take to make a human? According to a rough estimate calculated by the Thomas Jefferson National Accelerator Facility in Newport News, Virginia, a person who weighs 70 kg/154 pounds would have approximately *seven billion billion billion* atoms.[75] In scientific notation that is 7×10^{27}. Visualize a seven followed by twenty-seven zeros. But not just any atoms will do. A human body requires specific kinds of atoms in a specific ratio. There are six atomic elements that account for about 99% of a human's body mass. In an 80 kg/176 pound human body, the elements stack up as follows:

- **oxygen** (52kg, more than half of body mass)
- **carbon** (14.4 kg)
- **hydrogen** (8 kg)
- **nitrogen** (2.4 kg)
- **calcium** (1.1 kg)
- **phosphorus** (880 grams)
- **potassium** (200 g)
- **chlorine** (120 g)
- **sodium** (120 g)
- **magnesium** (40 g)
- **iron** (4.8 g)
- **fluorine** (3 g)
- **zinc** (2.6 g)
- **silicone** (1.6 g)
- **strontium** (0.37 g)
- **copper** (0.08 g)
- **manganese** (0.0136 g)
- **molybdenum** (0.0104 g)[76]

In order of their abundance (i.e., number of atoms) within a human body, hydrogen, the most common element in the universe, is also the most common element in the human body. Hydrogen atoms formed during the Big Bang, all others were forged inside of stars and then dispersed throughout the universe when those stars died and exploded in what's called a supernova event. Achieving live human status requires many specific kinds of atoms in just the right ratios, put together in just the right ways. So how do so many trillions of atoms come together to form a walking, talking human being?

A HUMAN BODY IS . . .

- **. . . a collection of atoms in flux.** People are constructed of approximately 7×10^{27} atoms.[77]
- **. . . 99% hydrogen, oxygen, carbon, and nitrogen atoms.** More than half of these are hydrogen, the most abundant element in the universe.
- **. . . mostly empty space.** People are built out of atoms and an atom is mostly empty space.
- **. . . mostly water.** Ratios vary by age, gender and weight, but water accounts for most of a person's bodyweight.[78]
- **. . . made up of tens of trillions of cells unique to homo sapiens.** A recent study estimates there are approximately thirty trillion such cells in an average human body, more than eighty percent of which are red blood cells.[79]
- **. . . a diverse habitat for many trillions of non-human cells.** Thirty to perhaps as many as three-hundred trillion non-human cells inhabit every human body.[80] Most of the microbes that live with us are unknown to science.
- **. . . supported by a bone framework.** At birth a human infant has about 300 bones, many of them fuse together, however, reducing the total number to 206 in adulthood.
- **. . . is bloody.** About seven percent (1.2 to 1.5 gallons or 4.5 to 5.5 liters) of an average person's weight is blood.[81]
- **. . . muscular.** There are approximately 700 named muscles in a human body.[82] There is no agreed-upon exact figure because some muscles are combinations of smaller muscles and can be interpreted differently.
- **. . . hairy.** People are covered by hair. The few hairless areas include lips, palms and soles of the feet.

Whether it is to make a rock, coffee table, or a person, atoms must first chemically bond to create molecules. With some exceptions, a molecule is the smallest unit of something that still retains the properties of that thing. For example, more than half of our bodies are water, and water is a specific molecule that consists of two hydrogen atoms bonding with one oxygen atom. This means there is a lot of hydrogen in every human. To make a human we go from atoms to molecules to cells. It is this transition from molecules to cells that is easily recognized as the emergence of life because cells function much like our bodies. Cells take in energy, digest food, excrete waste, take in oxygen, and reproduce. They are the fundamental unit of life. Some lifeforms do well with one cell (e.g., bacteria and protozoans) and others require a lot of them connected and working together. Among our estimated *thirty trillion cells*,[83] red blood cells are the most common type, making up more than eighty percent of a person's total cell count. But they are extremely small so they only account for about four percent of total body mass.

Seven billion billion billion atoms and thirty trillion cells may be fun numbers to toss about and contemplate because they're so big. But there's more, a lot more. Those gaudy figures don't account for the atoms, molecules, cells, and complex lifeforms that share the journey of life with every human.

I, COLONY

A significant part of the answer to the question *Who are we?* must take into account all the *other life* that lives on, in, and immediately around us. We cannot think of a human as a lifeform in isolation because we never have and could not exist without our colonists. We host trillions of other lifeforms and this is a fundamental to what it means to be human. Microbial life surrounds and permeates every human's existence.

Humans are superorganisms, much more than an arrangement of *Homo sapiens* cells alone. We are mobile, diverse, and thriving ecosystems. You are something like a rainforest with opposable thumbs, a coral reef in shoes. This is a profoundly important aspect of who we are. Every human is host to a wildlife park teeming with invisible lifeforms. We survive, tolerate, and depend on the moment-to-moment collaborations and conflicts of trillions of microscopic creatures. Yes, it's true, you are not quite the individual you may have been led to believe. Although their numbers are large and their impact significant, collectively they account for only about three percent or five pounds of a typical person's bodyweight.[84] What is most important to understand is that these creatures are not mere freeloaders or bystanders. They are active lifeforms, many of whom constantly do things with us and to us. Some are potentially harmful, some have neutral or unknown impacts, and others are good, essential even to our health and survival and more than their mere presence matters. Their numbers, ratios, and *their health* have direct implications for *our health*. Scientists are just now beginning to shed light on how numerous and important our microbes are to us. I recommend paying close attention to news that comes out of research on human-body microbes in the coming years because the more we learn about them, the more we will know ourselves.

THE ANCIENT INVADERS WHO STAYED

Viruses are incredibly prolific and effective parasites. Though they fail to qualify for most definitions of life, viruses have been busy for billions of years infecting, changing, hijacking, zombifying, killing, and generally having a profound influence on the evolution of life on Earth.[85] We tend to view them as an enemy because some make us sick, but most have other agendas and do not harm people. Enemy or not, however, it turns out that viruses are a lot closer to us than most realize.

Like modern people, our distant ancestors, dating back tens of millions of years, were intimately entangled with viruses and under constant assault from some. Some of these ancient battles ended in what might be described as a draw, with the virus blending in with their hosts and staying put. A result is that today some one hundred thousand bits, or roughly five to eight percent, of the human genome is made up of genetic material that came from ancient viruses.[86] Researchers are only at the beginning stages of figuring out if these invaders who stayed help, harm, or do nothing of consequence to us. Some suspect it is a complicated mix of many effects.[87] What we can be sure of today is that humankind is part virus. Or should we say some viruses part human?

Reflect for a moment about the universe of life you currently share your existence with. Your body is a universe of diverse habitats. Like Earth, you have your own versions of rainforests, deserts, mountain ranges, savannahs, lakes, and wetlands. Hair, folds of skin, warmer regions, cooler regions, dark places, sun-exposed places, they all host lifeforms evolved to live there. You even have your own atmosphere of sorts. It is the airborne fallout of your immense biodiversity. This living mist exists in the form of a unique cloud of microbial life swirling around you at all times.[88] A staggering diversity and abundance of lifeforms inhabit your ecosystems. Some of them are having a say in how well you can defend against disease or infection, what kinds of foods are best suited for you, how fat or thin you are, how well you will sleep tonight, and even what kind of mood you are in right now.[89] Perhaps the most remarkable thing of all regarding the resident microbes that help to define us, is that we know so little about them.

A majority, perhaps more than ninety percent, of the microbes we live and die with are unknown to science today.[90] They evolved with us over eons, are with us every moment of our lives, and yet they remain mostly mysterious. At least the work of science has revealed their presence to us and indicated that our personal microbes are of immense importance to us. For this we can be thankful. Any description of "human" must acknowledge these important strangers. Every person is a colony, a vast and complex collection of tangled, interdependent life. Our bond with them is such that these microbes are not so much riders, parasites, and assistants as part human, and we are part microbe.

A PLACE IN NATURE

Where do humans fit in with all the other life on Earth? Given the current immersion in concrete, plastic, and electronic environments many people know today it may be easy to ignore, forget, or never learn that we are a part of a greater natural world all around us. Fortunately, Carl Linnaeus, an eighteenth-century Swedish scientist came up with a helpful way of making some sense out of such an abundance and diversity of life. Linnaean taxonomy, or scientific classification, is a system that provides context and acknowledges similarities and kinships. It is often compared to Russian nesting dolls, with a large kingdom containing life forms with similar features subdivided in to increasingly smaller groups. Binomial nomenclature, commonly referred to as the "scientific" or "Latin" name, is the basic naming system for life forms. First comes the larger genus, followed by the smaller species. With *Homo sapiens*, for example, *Homo* is the genus, *sapiens* the species. This is Latin for "Wise Man". The life sciences have come far the last couple of centuries, of course. Never has life appeared more abundance and complex than today and the Linnaean system is inadequate to make sense of it all.[91] The genetic revolution has given scientists a vast resource of detailed information, and it doesn't always align well with classical taxonomy. Though imperfect, however, the Linnaean system remains useful and until replaced will continue to help scientists and laypersons alike sort and categorize the natural world.[92]

LINNAEAN CLASSIFICATION OF HUMANS

- **Domain:** Eukaryota (also includes fungi, tubeworms, redwood trees, whales);
- **Kingdom:** Animalia;
- **Phylum:** Chordata;
- **Subphylum:** Vertebrata;
- **Class:** Mammalia;
- **Subclass:** Theria;
- **Infraclass:** Eutheria;
- **Order:** Primates (also includes apes and monkeys);
- **Suborder:** Anthropedia;
- **Superfamily:** Hominoidae;
- **Family:** Hominidae;
- **Genus:** Homo (includes the Neanderthals, far-ranging *Homo erectus*, tiny *Homo floriensis*, and other extinct humans);
- **Species:** sapiens

If you don't want to keep the entire Linnaean breakdown for humans straight in your head, at least try to be familiar with these important classifications directly relevant to you: Domain: Eukaryota (you share this category of life with all animals, plants, and fungi); Subphylum: Vertebrata (you and all animals with a spine); Class: Mammalia (you and all endothermic animals with a spine, hair, mammary glands); Order: Primates (lemurs, monkeys, apes); Genus: Homo (this includes all forms of humans both alive and extinct); Species: Sapiens (modern humans). Being aware of these most basic groupings and associations can go a long way toward maintaining a realistic understanding of where you fit in with all the life around you.

WE ARE NOT ALONE

It is a common human experience to look up at the stars and wonder if we are alone in the universe. But while this remains an open question, we know we are not alone down here. We share the earth with many lifeforms, so many that we can only struggle to guess at their numbers. Known or unknown, our connection to this life is of profound importance to us all. A description of who we are must include due mention of the life we share this world with.

It is may be easy to believe that our increasingly high-tech and complex societies have moved us further away from nature. Trees, rivers, fungi, birds and bacteria may seem trivial if not altogether inconsequential to the life of a modern person in Lagos or London, but science tells us that the opposite is true. We are a part of nature and nature is central to our existence, whether we recognize it and act accordingly or not. Human activities are slashing global biodiversity and disrupting ecosystems everywhere today. But the surprising truth is that in the high-tech, plastic, concrete twenty-first century we rely on wildlife and natural processes not less so but at an *increasing* rate. Our growing population, modern economic systems, and complex social structures are making us more dependent on nature's services for our health, security, and prosperity.[93] We need nature.

In learning about the natural world, we discover vital things about who we are. Any explanation of humankind that does not include due attention to nature and our place in it is incomplete and inadequate, if not worthless. David George Haskell, an award-winning US biology professor, believes being aware of the connections and continuity of life helps us experience and understand humankind:

The interior quality of our minds is itself a great teacher of natural history. It is here that we learned that "nature" is not a separate place. We too are animals, primates with a rich ecological and evolutionary context. By paying attention, this inner animal can be watched at any time: our keen interest in fruits, meats, sugar, and salt; our obsession with social hierarchies, clans, and networks; our fascination with the aesthetics of human skin, hair, and bodily shapes; our incessant intellectual curiosity and ambition. . . . [W]atching ourselves and watching the world are not in opposition; by observing the forest, I have come to see myself more clearly. . . . Part of what we discover by observing ourselves is an affinity for the world around us. The desire to name, understand, and enjoy the rest of the community of life is part of our humanity.[94]

Everyone knows that living in cities can be stressful but is the urban environment making us crazy? It would seem so, at least to some degree. Despite their many benefits, there is something about modern cities that does not agree with us.[95] From noise to toxins, from lack of personal space to the loneliness of being a constant stranger in crowds, many reasons have been proposed to explain the problem of modern life. Still, specific causes for the correlation between the rise of urbanization and higher rates of mental illnesses are not clear. Could it be as simple as less nature equals more stress? Does more nature mean more peace and productivity for the human mind? There is a lot of complexity here but natural environments, the science strongly indicates, helps keep us mentally fit and feeling well.[96] They have been shown to alleviate mental fatigue which is thought to be a key factor in many outbursts of aggression and violence.[97] So, is green space the key to achieving world peace? I don't think it's quite that simple, but this certainly is relevant information for you and the quality of your life. When I walk around a major city and soak in the excitement of so many sights and sounds, I am surrounded by humans who may have been tamed but not domesticated. The wild stirs within us still. Grass, bushes, trees, lakes, rivers, the ocean, and clouds. Insects, birds, and giraffes. These things matter. They feel like home to us because for most of human existence they were.

Remember that a wilderness environment, or something close to it, can help your mind recover from hard work and the stress of daily life. Scientific research has revealed that it can make you feel better and perform better. A recent and notable Stanford University project compared what ninety minutes spent walking in a natural landscape did to the brain compared to ninety minutes walking in an American city.[98] The results may not have been surprising, but they were remarkable, nonetheless. A simple walk among trees, bushes, and birds significantly reduced self-reported rumination, the negative and obsessive thought patterns linked to depression and other mental health problems. There also was less neural activity in the subgenual prefrontal cortex. This is an area of the brain associated with social withdrawal in both healthy and depressed people. The test subjects who walked in the city environment did not experience these positive effects. "This study," write the researchers, "reveals a pathway by which nature experience may improve mental well-being and suggests that accessible natural areas within urban contexts may be a critical resource for mental health in our rapidly urbanizing world."[99] Green is good for the human mind.

The personal and practical aspects of your deep and intimate connection to wildlife and ecosystems are numerous and well supported by scientific research.[100] Studies have shown, for example, that living in or within sight of a natural landscape can boost and help maintain good mental health.[101,102] There is excellent evidence that shows living in or near wild environments can make us healthier[103] and lengthen lifespans.[104] All of this makes sense, of course, because we are of the wild. The natural world made us who we are over an overwhelming majority of our existence. It should be no surprise, then, that nature still calls to us.

CLIMBING DOWN AND STANDING UP

A common misconception circulated by many people today is that scientists claim modern humans evolved from modern apes and monkeys. This is not true, of course, and unnecessarily confuses many people. The current claim by science, based on strong anatomical, fossil, and genetic evidence, is that we are very closely related to modern great apes, the chimpanzees most of all.

We share a *common ancestor* with chimps that lived millions of years ago. It was widely thought that the human line split with the chimp line about six million years ago in Africa.[105] In recent years, however, some researchers have found evidence that seems to push the split back to somewhere between ten and thirteen million years ago.[106] It is difficult and might prove impossible to pinpoint a specific time for the last common chimp-human ancestor because it may be that our evolutionary lines didn't split just once over a short period. Our process of separation may have been long, tangled, and inconveniently fuzzy for modern analysis.[107] While it is fascinating and important for us to try and learn when this happened, what is perhaps most important for you to know right now is that modern apes are close kin and that they and their ancestors feature prominently in our story.

BONOBO COUSIN. Modern humans did not evolve from modern apes, obviously. *Homo sapiens* does, however, share common ancestors with the living great apes. Some 99 percent of human DNA is identical with *Pan paniscus*, commonly known as the bonobos. *Photo by the author*

Talk of chimp-human kinship may be unsettling to some people but having animals such as bonobos in our extended family need not be something to cringe at. I have studied bonobos for years and spent considerable time observing them at the San Diego Zoo. I can assure you that these are remarkably intelligent creatures who are among the most elegant, complex, and peaceful lifeforms on this planet. If anything, we might accept our evolutionary proximity to them as a source of some pride. It is possible that we might even stand to learn some things of value from them.

THE HOMINID FAMILY

Current living members by genus:

- *Gorilla*: gorillas, Western Gorilla (*Gorilla gorilla*) and Eastern Gorilla (*Gorilla beringei*); Gorillas and humans share about 98 percent of their DNA.[108]
- *Homo*: *Homo sapiens*, humans.
- *Pan*: Bonobos (Pan paniscus) and the common Chimpanzee (Pan troglodytes). Chimpanzees and humans share 99.6 percent of their DNA.[109]
- *Pongo*: orangutans Bornean orangutan (P. pygmaeus) and the Sumatran orangutan (P. abelii); Orangutans and humans share about 97 of their DNA.[110]

Because we are primates no one should be surprised that our children love to climb trees and swing from branches. They come by it naturally. I still haven't outgrown the impulse to scramble up a really good climbing tree when I see one. Culture—our swirling and ceaseless hurricane of technology, taboos, beliefs and behaviors— may lead us to feel more distant from our primate kin than we are but the close anatomical and genetic relationship we share is clear. Even the human brain, a super-organ that makes possible poetry and lunar landings, is something close to a scaled-up, neuron- bloated chimp brain. Our genetic kinship to apes, chimpanzees in particular, is striking as well. We share more ninety-nine percent of our DNA with modern bonobo chimpanzees (*Pan paniscus*) and common chimpanzees (*Pan troglodytes*).[111] This may be somewhat misleading, however, because we know little or nothing about the function and importance of many human and chimp genes. But it does give us a reliable indicator of how relatively close we are to an orangutan compared to a bird or ant, for example. One fascinating thing that most people probably don't know is that chimps, gorillas, and orangutans seem to be more closely related to us than they are to each other.[112]

Some people take unnecessary offense to the close relationship we share with other primates. This is unfortunate because we can find value in accepting and exploring the reality of who we are. Humans are not chimpanzees, but the behavior of chimps can offer us insights about ourselves and clues about some aspects of how we may have lived long ago. Perhaps even more important is that the link between them and us can open our eyes to the larger connections we share with all other animals and all life, now and in the past. Acknowledging our evolutionary proximity to chimps and other animals does not, however, give us permission to behave as savages forever free of guilt.

Our close physical, genetic, and behavioral relationship to modern chimpanzees does not diminish our humanity. It can do precisely the opposite for those who choose to look at this in a different light. Knowing as much as we can about who we are, where we come from, and which lifeforms we are most closely related to lays bare the awesome value of a remarkable journey and our potential as intelligent, creative beings. Seeing the great distances that we have covered in relatively short time can help us to believe in how far we might go.

NEANDERTHALS, HOBBITS, AND US

We belong to a brilliant, boisterous, and beautiful family. It includes all great apes, both living and extinct. Our specific genus, *Homo*, contains all human species, and there have been many. But we, *Homo sapiens*, are the only one left alive. All other human species are extinct. The earliest members of our genus date to around 2.5 million years ago when humans evolved from more chimp-like ancestors called *Australopithecines*. They walked upright but had much smaller brains than us.

It is important that we do not imagine human evolution as a neat and orderly path. Our ancestors did not systematically upgrade, one species at a time, with each new model introducing improved features such as a bigger brain. It was more complicated than that. Several ape species evolved over millions of years what we might describe as more-human-like traits but all of them succumbed to extinction anyway—except for one. We are the last humans standing. The winding, convoluted path that led to us is why the human story remains incomplete today. We rely on difficult-to-find fossils to help us piece together our past. This is a profoundly difficult challenge when we have so many prehistoric actors to place in proper context.

OUR CROWDED GENUS

➢ *Homo habilis* ("Handy Man")
Time: 2.4 million to 1.4 million years ago
Range: Eastern and Southern Africa
Height: 3 ft. 4 in. - 4 ft. 5 in. (100 - 135 cm)
Weight: average 70 lbs. (32 kg.)
Notes: *Homo habilis* ("Handy Man") may be the oldest member of our genus. They evolved from earlier, more chimpanzee-like Australopithecines and were discovered at Olduvai Gorge by a team led by famous fossil hunters Louis and Mary Leakey. They were given the name because of their association with thousands of stone tools found in the same location and context.[113]

➢ *Homo erectus* ("Upright Man")**Time:** 1.89 million to 143,000 years ago
Range: Northern, Eastern, and Southern Africa; Western Asia (Dmanisi, Republic of Georgia); East Asia (China and Indonesia)
Height: from 4 ft. 9 in. - 6 ft. 1 in. (145 - 185 cm.)
Weight: from 88 - 150 lbs. (40 - 68 kg.)
Notes: An extremely successful species lasting approximately two million years. *Homo erectus* were the first people believed to have controlled and used fire; first people to migrate out of Africa and survive in diverse environments around the world.[114]

➢ *Homo naledi* ("Naledi" is reference to the Rising Star caves system where the fossils were found)
Time: 335,000 - 200,000 years ago

Range: southern Africa
Height: 5 ft (150 cm)
Weight: 100 pounds (45 kg)
Notes: It was one of the most stunning discoveries ever: Some 1,500 fossils representing fifteen individuals were found in the Rising Star Cave system near Johannesburg in South Africa (2013*)*. Some paleoanthropologists think the bodies may have been placed in the cave as ritualistic behavior indicating complex culture, but this is not certain. *Homo naledi* is a fascinating combination of physical features. Its hands and feet are more like those of a modern human species than the earlier Australopithecines. Brain volume, however, is only about 40 percent of modern humans, more in line with Australopithecines. *Homo naledi* might be the ideal species to remind us that our evolutionary story is complex, blurry, and cannot be described in the simple step-by-step manner some may prefer.[115]

➢ *Homo heidelbergensis* (Named for Heidelberg, Germany where fossils were found)
Time: 700,000 to 200,000 years ago
Range: Europe; Asia (China); Africa (eastern and southern)
Height: Males: 5 ft. 9 in. (175 cm.); Females: 5 ft. 2 in. (157 cm.)
Weight: Males: 136 lbs. (62 kg.); Females: 112 lbs. (51 kg.)
Notes: First people known to live in colder climates and first people known to have controlled and used fire. First people known to have hunted very large animals. Built shelters from wood and stones.[116]

➢ **Denisovans**
Time: 100,000 years ago
Range: Asia and Europe?
Height: unknown
Weight: unknown
Notes: The discovery of a tiny finger bone in a Siberian cave turned out to hold big surprises when researchers successfully extracted nuclear DNA from the bone and found it to be from a previously unknown member of our genus. The Denisovans were perhaps a cousin or Asian version of the Neanderthals. Genetic analysis also revealed that the Denisovans had mated with both Neanderthals and anatomically modern humans

over the last 100,000 years or so. Some modern people today carry genes from them as well. The Denisovans hold a unique distinction as the only extinct people to have been identified by genetics rather than fossils. [117]

➢ *Homo floresiensis* (Nicknamed "hobbits" for their diminutive size)
Time: 100,000 – 50,000 years ago
Range: Asia (Indonesia)
Height: 3 ft. 6 in. (106 cm.)
Weight: 66 lbs. (30 kg.)
Notes: Made stone tools and hunted small elephants. May have been driven to extinction by a volcanic eruption. Nicknamed "Hobbits".[118]

➢ *Homo neanderthalensis* (Named for Neandertal valley in Germany were first fossils were found)
Time: 400,000 - 40,000 years ago
Range: Europe and southwestern to central Asia
Height: Males: 5 ft. 5 in. (164 cm); Females: 5 ft. 1 in. (155 cm.)
Weight: Males: 143 lbs. (65 kg.); Females: 119 lbs (54 kg.)
Notes: Closest extinct relative to modern humans. Neanderthals coexisted with modern humans as recently as 40,000 years ago. Unknown at this time if they went extinct due to environmental changes or were outcompeted or violently driven to extinction by anatomically modern humans. Neanderthal brains were as big or larger as modern human brains. They had relatively short and muscular bodies. Neanderthals had complex cultures and left behind a genetic legacy. Nearly all people today who do not have relatively recent African ancestry carry Neanderthal genes in their DNA.[119]

➢ *Homo sapiens* ("Thinking Man")
Time: Evolved in Africa approximately 200,000 to 300,000 years ago. Still alive with a current population of more than seven billion individuals.
Range: Dispersed widely, lives in almost all land environments, including a small presence in Antarctica and even space (International Space Station) where a small transient population has lived continuously since November 2000.
Height: Average adult 5 ft. 5 in. (170 cm.)[120]

Weight: Average adult: 136.7 lbs. (62.0 kg.)[121]
Notes: Modern humans have relatively large brains which are, on average, approximately 1300 cubic centimeters. The species is notable for remarkable intelligence-based abilities, including: science, spaceflight, music, complex oral and written language, mathematics, creative fiction, etc. Humans currently are degrading the Earth's biodiversity at a rapid rate via destruction of natural habitats, over hunting/fishing, pollution, spreading invasive species. *Homo sapiens* is capable of self-inflicted extinction by technological means (Nuclear war, for example).[122]

One of the last human species we shared the planet with was *Homo floresiensis*. Nicknamed "hobbits" because of their diminutive stature, these roughly one meter or three-and-half-feet tall people with grapefruit-sized skulls lived in Indonesia until around 50,000 or 60,000 years ago.[123] Neanderthals went extinct even more recently, about 40,000 years ago.[124] The oldest fossils of anatomically modern humans, people who were physically indistinguishable from us, date to from 200,000 to 300,000 years ago. If these dates hold up to future discoveries it means that we have been alone as the sole human species on Earth for only around thirteen to twenty percent of our existence. Most people, I suspect, are unaware of this and would find it surprising. But it is a significant, fundamental fact about who we are. We live today as one successful (so far) version of many human species. Experts speculate endlessly about whether or not our direct ancestors merely outlasted or won direct competitions for resources against the Neanderthals and Erectus populations they bumped up against. Some suggest that our ancestors may have wiped out all rivals by violent means, as in prehistoric genocide.[125] We may never know exactly what happened, but I suspect it was some combination of all of the above.

THINKERS, TALKERS, PLANNERS, AND DREAMERS

There is a lot more to us than bipedalism and big, complex brains. We didn't end up as tragic prehistoric cat food or being the victims of any number of possible evolutionary dead-ends because we thought, imagined, and planned. We *cultured* our way through one big challenge after another. There may have been some random luck involved along the way, of course, but we have plenty of reasons to be proud of our ancestors. Life in the prehistoric world was no picnic for such relatively small, slow, and weak primates. Consider the horrifying predators that coexisted with us. These included the *Pachycrocuta, a monstrous hyena the size of a modern lion;* large and stealthy saber-toothed cats; and even giant birds capable of snatching and airlifting smaller hominins away to their deaths with powerful talons. There also were diseases, storms, wildfires, earthquakes, volcano eruptions, and climate changes to contend with. Given our comparative lack of strength and speed, tiny teeth and thin skin, it is extraordinary that we have survived and thrived as we did. Cleary our brains were at the center of this success story. We were and are, more than anything else, thinkers. Brainpower has always been our superpower. Before I praise the intellectual greatness of humans any further, however, it must be stated that we are also incredibly stupid. Our impressive and unprecedented intelligence is often applied toward thoughtless and destructive behaviors that harm not only ourselves but much of the life around us too. Therefore, let's not make the mistake here of confusing intelligence with wisdom or being smart with being sensible. We may be the most intelligent creatures on Earth but it's probably fair to say that we have not yet lived up to that name we gave ourselves, *Homo sapiens* or "wise man".

Some experts believe that a fundamental change in our thinking occurred relatively suddenly around 30,000 to 50,000 years ago.[126] They base this on the hard evidence of abstract thought and complex social behaviors that shows up in the archaeological record, mostly in Eurasia. Other experts, however, point out that much earlier artifacts, left behind by African populations, indicate that the same kind of "modern" thinking was already occurring more than 100,000 years ago.[127] For this new way of using brains to arise over two or three hundred thousand years would still be considered quick and revolutionary in geologic terms, but it's not so sudden or explosive within the context of human evolution. Keep in mind the challenges of archaeological work, too. The further back we go in time the more difficult it is to find evidence of human activity. For example, it is likely that our ancestors used wood tools before and more frequently than stone tools because wood is easier to shape into implements. But it rots rapidly and usually is lost over time so, most likely, archaeologists have had to rely disproportionately on stone artifacts to inform our ideas about how crafty and savvy the earliest people were. I'm not suggesting that we can pretend to know them with imagined artifacts but it's likely that an unfortunate stone preservation-and-discovery bias distorts our view. In general, be cautious about overdramatizing the human or cognitive revolution because the rise of consciousness and culturally complex behavior appears to have happened much earlier and been more gradual than was once thought.[128] Allison Brooks, professor of anthropology at George Washington University, has been making this case for years. She contends that our intellectual upgrades were well underway in Africa much earlier than 50,000 years ago. She points to early evidence that has been discovered in Africa indicating that our ancestors were thinking and behaving in "modern" ways by the time we had evolved anatomically modern bodies, at least more than 100,000 years ago. Brooks and Sally Mcbrearty, a professor of anthropology at the University of Connecticut, co-authored a paper that summarized this position. They write in the *Journal of Human Evolution*:

> [T]he "human revolution" model creates a time lag between the appearance of anatomical modernity and perceived behavioral modernity, and creates the impression that the earliest modern Africans were behaviorally primitive. This view of events stems from a profound Eurocentric bias and a failure to appreciate the depth and breadth of the African archaeological record. In fact, many of the components of the "human revolution" claimed to appear at 40-50 ka [thousands of years ago] are found in the African Middle Stone Age tens of thousands of years earlier. These features include blade and microlithic technology, bone tools, increased geographic range, specialized hunting, the use of aquatic resources, long distance trade, systematic processing and use of pigment, and art and decoration. These items do not occur suddenly together as predicted by the "human revolution" model, but at sites that are widely separated in space and time. This suggests a gradual assembling of the package of modern human behaviors in Africa, and its later export to other regions of the Old World.[129]

HUMANS IN BRIEF

- **Population:** More than 7.6 billion; Currently projected to reach 9.8 billion by 2050 and 11.2 billion by 2100.[130]
- **Origin of** *Homo* **genus:** *Homo habilis* is currently thought to be the earliest human species, the first and oldest member of the *Homo* genus. These people date back to about 2.5 million years ago. *Habilis* was immediately preceded by the more chimpanzee-like Australopithecines. *Homo erectus* evolved from habilis around two million years ago and would eventually evolve into modern humans. *Homo sapiens* first appeared in Africa 200,000 to perhaps 300,000 years ago and expanded throughout Africa and beyond that continent in multiple waves or a single wave from 120,000 to 60,000 years ago.[131] Fossils found in Jebel Irhoud, Morocco seem to indicate that anatomically modern humans were living in Africa as early as 315,000 years ago.[132] Always keep in mind that the human story is still being written. New fossils and new analysis of old fossils continue to present potential plot changes. For example, what was thought to be a *Homo erectus* skull discovered in China's Shaanxi Province back in 1978 is now being reconsidered as possibly a *Homo sapiens* specimen.[133] If so, its 260,000-year-old age and location far from Africa could mean a profound shakeup for the Out of Africa origin theory that best fits with the current overall fossil and genetic evidence.
- **Keys to survival:** Humans have relied on social networks, language, tools, and culture to survive diverse environments around the world.[134] Other lifeforms (e.g., woodpecker finches, sea otters, wild apes, monkeys, and more) use tools, but no living creature on Earth comes close to relying on technology as much as humans do. Communication also is of vital importance. It enables people to plan, share discoveries, build on ideas, make predictions, and much more. There were 7,099 known languages in 2017.[135] Information sharing continues to hold immense importance for our species today. More than half of the current human population (3.8 billion) is online and approximately forty percent of all people (three billion) use at least one social media platform.[136] More than two billion are on Facebook alone.[137]

- **Diet:** Humans were omnivorous hunter-gatherers, living on a wide range of plant and animal foods, for more than 99 percent of the more than two-million-year existence of the Homo genus. Dietary flexibility was a significant survival advantage in prehistory. Anatomically modern humans were hunters and gatherers for about 95 percent of their 200,000-year existence. Today virtually all humans rely on agriculture, pastoralism (raising various animals for meat, milk, eggs, clothing), and/or fishing. To date, people have converted or cleared more than thirty-five percent of the Earth's ice-free land surface for farming and pastoralism/grazing.[138] Agriculture today uses more than sixty times more land than all cities and suburbs combined.[139]
- **Gestation period:** 280 days; human infants are born virtually helpless and remain dependent on parents and/or other adult humans for years.
- **Brain size:** Approximately three pounds (1.5 kg.); volume range is around 1100 cubic centimeters for female and 1400 cubic centimeters for males; a typical brain contains approximately 86 billion neurons[140].
- **Human universals:** There are many, including: oral language, fire, cooking food, family units, use of names for people, weapons, magical thinking, rites of passage, body adornment among them.[141] Humans are also heavy daydreamers, spending nearly half their waking hours doing it, according to one study.[142]
- **Kinships:** Modern humans are very closely related to one another, having *more than ninety-nine percent* identical DNA. Modern chimpanzees and modern people are closely related as well, sharing approximately ninety-eight percent identical DNA. Chimps are more closely related to humans than they are to gorillas. The last common ancestor of chimps and humans probably lived between six to thirteen million years ago.[143]
- **Life span:** The current global average human lifespan is 70 years (varies widely by society). According to a United Nations study, life expectancy for humans in Africa is 60 years; 72 years in Asia; 75 years in Latin America and the Caribbean; 77 years in Europe and in Oceania (Pacific region); and 79 years in Northern America.[144] The United Nations projects global life expectancy for all people to rise to 77 years in

2045-2050 and to 83 years by the end of the 21st century.

- **Conservation status:** Humans are designated "least concern" by the International Union for Conservation of Nature (IUCN).[145] Humans are thriving at present with a widely dispersed population rapidly approaching eight billion.
- **Threats.** Possible *Homo sapiens* extinction events, with varying or unknown degrees of likelihood and danger, include: nuclear war; major asteroid strike; supervolcano eruption; natural or engineered viral pandemic; global ecosystem collapse; gamma-ray burst from a star; malevolent use or loss of control of nanotechnology or artificial intelligence.
- **Known extinct humans:** *Homo habilis, Homo naledi, Homo erectus, Homo floresiensis, Homo heidelbergensis, Homo Neanderthalensis.*

Regardless of timing and location, our great cognitive transition was momentous. This shift to a reliance on more and more complex thinking as well as increasing cultural complexity is central to the question of who we are now. It was long ago in prehistory that we became true thinkers and dreamers. Language, abstract symbols, art, rituals, spirits, and gods became vital to hunter gatherers and the first farmers many thousands of years ago. It was not enough for us to live in the world; we sought to explain and give it meaning, too. In figuring out their environments for survival—while also embellishing, imagining, misperceiving and fantasizing about the reality around them—a few bands and tribes steered a course to the present we currently live in. Do not overlook the obvious. Our modern societies have been built not only by lumber, steel, and concrete, but also by countless agreed-upon mythologies and purely imaginary ideas. Governments, supernatural beliefs, money systems, nations, and corporations are examples of things people made up and collectively agreed to accept as real. This began a long time ago, initiated by people most of us today rarely if ever think about. So much of our reality is not rooted in reality, and prehistoric people are the reason. It is important to think about this because it's a significant part of who we are. We are a species with a strong penchant for imagining and believing.

It is difficult to overstate the significance of intelligence in our story. Take away intelligence, remove our capacity and compulsion for imagination and planning, and we are but taller chimpanzees. The human brain was our secret weapon. What does it matter if a lion is faster or a mastodon stronger? We defeated the lion and the mastodon by *thinking*. Our ancestors went from prey to predator, from scavenger to master by analyzing threats, talking about them, sharing ideas with others, and then trying those ideas out. From human brains came the stone hand axe, spear, torch, fish hook, animal trap, atlatl, boomerang, bow-and-arrow, and complex hunting strategies.

Humans possess an inherited knack for creative problem solving. The talent is within you; it's a human trait in all of us. Unfortunately, we in modern societies tend to get so caught up in finite degrees of ability and precisely measured performance that we overlook our common, nearly universal wonders. Sure, some people may be profoundly gifted. Some might be better suited for graduate level physics classes or the Olympic marathon than most. But virtually all of us can think about the atom, the universe, and at least glimpse at how everything works. Virtually all of us can run or walk many miles, too—or could if we lived a lifestyle closer to who we are meant to be according to our story and evolutionary inheritances.

The people who preceded us reasoned, planned, imagined, invented, and calculated their way to evolutionary success (so far). The manner by which they survived is the reason that today we can build a 450-ton, football-field-sized artificial habitat in space, communicate near instantly across great distances, and write beautiful poetry. Believing in a pseudoscience like astrology is possible for the same reason that doing astronomy is possible. A human newborn does not enter the world preloaded with all or most of the information necessary for its survival. We are unique in this regard. To survive, in prehistory or now, a human must learn a tremendous amount of information over a lifetime. Only a brain like ours, a malleable, plastic organ packed with many tens of billions of shifting neurons engaged in endless networking can handle this demand.

When we transitioned to a species that relies so heavily on cognitive abilities, we became the most powerful and profoundly weird creatures of all time. More than 7.5 billion people carry in their heads a three-pound blob of strange magic, an electrochemical storm of genius and madness unprecedented and unsurpassed in this planet's four-billion-years of natural history. This is who humans were. This is who *you are*. You are one more unique link in a long chain of fantastic inventiveness and brilliant imagination.

It is exciting to imagine the amazing changes that must have been going on inside our ancestors' craniums during the rise of complex culture, more effective technology, and the exploration of new environments. Flexibility, the ability to learn and change, is the brain's most valuable quality and it blossomed in full over the last few hundred thousand years. Millions of interconnected neural networks became the dense and tangled labyrinths necessary for the massive cognitive firepower of our species. We can imagine this because brains don't fossilize. Richard G. Klein, a professor of anthropology and biology at Stanford University, writes: "Arguably, the development of modern behavior depended on a neural change broadly like those that accompanied yet earlier archaeologically detectable behavioral advances. This explanation is problematic, however, because the putative change was in brain organization, not size, and fossil skulls provide little or no secure evidence for brain structure."[146] Learning via complex language within cultural settings that outlive individuals is a key factor in our becoming such a spectacular and impactful species. When our brains evolved into full-blown learning and communication machines, we began a process of sharing, saving, and transferring knowledge forward that continues today. This "ratchet effect"[147] is crucial to understanding humankind's rapid technological progress. Our reliance on brainpower and teamwork to survive in difficult prehistoric environments meant we had to prioritize teaching, socialization, and conformity. This allowed for the cultural accumulation of knowledge on a massive scale, a unique feature of humans. When information and skills are routinely passed on from older to younger generations it means people don't have to discover, figure out, and invent the same things again and again. They instead build on the best ideas and technology to make them better.

SEARCHERS AND WANDERERS

For most of prehistory various human species, including us, relied on a strategy of diversification. It was the "eat everything edible, whenever possible" diet. And it worked. They foraged for plants, fruits, and berries; swiped eggs from nests; and killed animals. They also scavenged and stole the kills of other predators when possible. If there was an abundance of food, the band stayed put. If food was scarce, it moved on. This survival strategy was dependent upon intelligence and endurance. It took a sharp mind to observe, learn and remember massive amounts of vital information about so many food sources. Walking and running many miles daily on average required extreme endurance capabilities. More ground covered meant more food collected.

We need to include the hunting and gathering lifestyle in the human definition because this is who we were for more than ninety-nine percent of our existence. And it is still who are to a significant degree. Farming and pastoralism only took off about five to ten thousand years ago. And urbanization, the phenomenon of most of the human population living in city environments is very new.

With an omnivorous diet and increasingly clever brains, members of our genus earned a place among the most opportunistic survivors of all time. Early humans were close to the land and wildlife around them. Crucial life-and-death knowledge for every band and tribe included the rhythms of weather and climate. They would have to have known where certain important plants grew and what time of year was best to eat them, the timing and location of animal migrations, the location of fresh water sources, and so on. They had to be fast learners, sharers of information, and creative in their approach to figuring out new ways to survive in new or changing environments.

Everyone knew everyone within early hunter-gatherer social units. Personal privacy would have been rare or nonexistent in prehistory. It is fascinating that in the early twenty-first century the rise of the Internet, the social media phenomenon, and the harvesting of personal data by businesses is rapidly rewriting the rules and expectations of personal privacy. An unexpected consequence of modern technology may be to return us back to prehistoric privacy norms. We don't know what the size of a typical hunter-gatherer group would have been. It probably varied widely. Studies of contemporary hunter-gatherer groups suggest that some prehistoric bands/tribes could have had as many a thousand or more members while a typical number would have been 77.[148] Based on what we know of modern hunter-gatherer bands, it is likely that no one would have owned land or accumulated much if anything in the way of personal property throughout almost all of prehistory. The human way for virtually all of our existence seems to have been to be smart, highly mobile, socially connected and interdependent, and profoundly resourceful. This was how we lived for more than two million years.

Roaming the land on an endless quest for the next meal may seem unreasonably harsh to most people today but it certainly was a successful strategy in prehistory. *Homo erectus* and later versions of humanity spread throughout Africa and then much of the world, not as farmers and goat herders but as hunters and gatherers. Given the long duration of this survival strategy (more than ninety-nine percent of our existence) and the impressive geographic expansion it enabled (Africa, Europe, Asia, Australia, the Americas), we ought to be aware of it and include it in our perceptions and descriptions of *Homo sapiens*. Sometime when you feel that itch of deep curiosity, maybe a nagging whisper of ambition, or the need to just get up off the couch and do something, you might recall the lives of your prehistoric ancestors, for they surely felt it too. They must have, because the urge to move, learn, and explore helped fuel a survival strategy that served them well. Perhaps we ought to consciously value and nurture these impulses when we experience them because they are a large part of what makes us human.

MOVING IS LIVING

Knowing how our ancestors survived for many hundreds of thousands of years is directly relevant to us today. Understanding that the modern human body evolved through a selective process that placed a positive survival value on walking, running, and thinking can inform how we might live our best today. Foraging for food on foot all day is not possible, of course, but it's a mistake to ignore what the past made of us. We have bodies and brains that did not evolve to sit still all day. The sedentary lifestyle common to so many modern people is extremely unhealthy for both the body and brain. Sitting all day in front of computer screens at work, followed by more sitting in front of a television at home in the evening often leads to severe health consequences, including premature death. But this only makes sense because it is a corruption of our genetic programming that was written over millions of years. In recent years, scientists have discovered that this is such a problem that even consistent exercise is not enough to compensate for the detrimental effects of prolonged sitting, which include increased risk of heart disease, diabetes, and cancers.[149] Anatomically modern humans did not sit ten to fifteen hours per day. They moved. So, considering we are working with very similar brains and bodies. Perhaps we should too.

WANT TO BE MORE CREATIVE?

For humans, walking is more than exercise or a way to get from point A to point B. It alters the mind in ways that can enhance creativity. An important Stanford University study found that 81% of test subjects showed significant elevations in creative thought (as measured by a specific test for it) while walking on a treadmill versus being seated. The researchers tested people in a variety of states: seated, walking on a treadmill, rolled in a wheelchair outside, and walking outside. Walking outside had the most positive impact on creativity. And the gains do not only occur while walking. Test subjects exhibited a "residual creative boost" after they had finished a brief walk.[150]

The scientists believe that walking has this effect because it loosens restraints and filters that executive function (focusing on a task) has on memory. When one dedicates mental effort to something the mind tries to help the person by shutting down what it deems to be irrelevant distractions from within. Walking seems to relax inhibitions that suppress memories that are not obviously or directly related to the task.[151] But to be creative, we want incongruent memories and disparate bits of information to flow because they are often the source of new and unique ideas. Walking opens the mind.

Those who cringe at thoughts of running marathons, or even walking around the block, it may help to know that a typical modern human body is an elite endurance machine—or at least can be. The human brain, in form and function, is meant to sit atop a mobile platform, one that stands, walks, and runs far more than is common for most people today.[152] Our past made us this way. We ran to live. For example, experts believe that "persistence hunting" was a common method people relied on to obtain meat for many hundreds of thousands of years.[153] The strategy requires people to chase an animal, a gazelle or antelope, for example, and keep it moving until it succumbs to exhaustion and offers a relatively easy kill. Such a chase can take many hours and cover many miles. This form of hunting might have even spurred or accelerated increases in brain power, some experts suggest.[154] Imagine running for hours, across often unfamiliar lands, constantly scanning for threats and opportunities, as you track a tiring animal. Sharper brains, those that excelled at noticing and remembering helpful details in the environment, would likely do better at this and be selected for over many generations.

We may think of ourselves as slow compared with many other animals but this is true only over very short distances. We excel at walking and running far. As endurance athletes, humankind is elite in the animal kingdom. Over a long enough distance, a good human runner can beat a horse, dog, lion, bear, and many other animals. Even the world's fastest land animal, the cheetah, can't keep up with an average human runner over distances beyond a half-mile or so. "The genome of *Homo sapiens* has evolved to support the svelte phenotype of an endurance runner, setting him/her apart from all other primates," concludes Mark Mattson, professor of neuroscience at Johns Hopkins University.[155] In recent years, Mattson and others have revealed that there are numerous positive brain benefits of endurance activity. A brain that is kept upright and on the move often is likely to function at a higher level and last longer than one trapped inside a sedentary body.[156] It's difficult to overstate the importance of this. Vigorous physical activity stimulates the growth of new neurons in the human brain throughout life.[157] This is a remarkable and recent discovery that most people are still not aware of: *Exercise grows your brain.* Moving is good for us in numerous ways. For example, studies indicate that running and walking reduces the risk of kidney cancer[158] and brain cancer mortality.[159] We are movers. Ignorance or denial of this can not only hold us back from optimizing our lives, it may lead us to premature death.

Cardiovascular diseases are the world's leading cause of death, claiming nearly eighteen million people each year.[160] This is *31 percent* of all deaths globally and much of the problem comes down to insufficient physical activity. Insufficient physical activity is now the fourth leading risk factor for mortality worldwide, according to the World Health Organization.[161] Based on research with contemporary hunter-gatherer societies, it seems certain that our ancestors were far from sedentary, neither were they burdened with high rates of cardiovascular diseases.[162] As important and well-established as the need for a daily jog, bike ride, swim, or walk is, it's important to be clear that exercise alone won't eradiate the problem of cardiovascular diseases. Smoking, poor nutrition, excessive eating, and environmental toxins have a lot to with the current high rates, too. Research on daily energy expenditures among modern hunter-gatherers, for example, indicates that the benefits of exercise can be effectively cancelled out by poor nutrition or excessive eating.[163]

Humankind spent most of its existence on the move, forever in search of food and safety.[164] And then we just stopped. The shift to agriculture and pastoralism led in turn to the very recent rise of urbanization. Most people today, fifty-four percent, live in cities and by the year 2050 sixty-six percent of the world's population are projected to live in urban environments.[165] Evolutionarily speaking, this is a trend that took off and transformed human lifestyles within the blink of an eye. Don't take this lightly because it occurred thousands of years ago. Today we both benefit and suffer, all of us, from this radical and relatively rapid change. Over many generations and without conscious purpose, humankind rebelled against its genetic programming and oldest traditions of mobility and deep, universal knowledge of the natural world. We sat down and stayed put, drastically reduced the diversity of our diets, staked our future to a few plant and animal species, and began the wild ride we call civilization. In many ways, we live our lives in defiance of who we really are and how our bodies and brains are built to function.

Relatively few people today are anywhere near as consistently physically mobile and active as our human ancestors were over more than two million years. It is unfortunate that many people today think of exercise first in the context of physical attractiveness. Vigorous exercise offers us something in addition to improving or maintaining health that is very important. A tough run or hike on a mountain or forest trail, for example, is a very *human* experience. It is the kind of thing that our direct ancestors did across thousands of generations every day for their survival. A long leisurely walk or good hard run can be thought of as an expression of one's humanity. There is a familiarity and an affinity for self-locomotion that remains in us—a need perhaps. You may not want to run or walk several miles per day. Some people can't for various reasons. Nevertheless, as a human you should know that you are meant to be a mobile animal. You belong to a *running and walking* species. Birds fly, fish swim, and humans walk. Every cell in your body contains DNA that harbors the genetic legacy of elite endurance athletes.

As something of a lifelong fitness fanatic, I have sometimes wondered why the next workout seems so crucial to my wellbeing. Deep down, something goes wrong for me if I go more than a few days without running, weight training, or dragging myself up some mountain. Inevitably my body and mind slip into a sluggish, wretched state when I'm inactive for too long. My mood darkens and I'm pretty sure that my cognitive abilities suffer as well. Not all of my runs and gym workouts have been great, but I can't remember ever regretting a single one. I don't just enjoy exercise; I need it. Perhaps this is merely some run-of-the-mill addiction. I could be hooked on the fleeting high of exercise-induced endorphins and nothing more. It's also possible, however, that I'm hooked on being human. I suspect the answer to why I walk, run, and labor under heavy weights may have something to do with feeling alive and being true to my existence. Moving, struggling, and sweating is one key to being a human in full.

So, *who are we?* Taking the long view, based on what we know of our evolutionary past, we are lifeforms with bodies and brains evolved to walk and run in natural environments, eat highly diverse diets, and overcome novel challenges. The past shaped our brains to excel against changing demands. We improvise and adapt, make and exploit discoveries, and learn new skills. Humans are extremely social, too, with a deep and constant desire to cooperate with others. All this suggests that it is time for our species to once again get up on its feet and start taking long walks toward exciting and endless horizons, both real and imagined.

SO ALIKE, SO DIFFERENT

> **Underneath the surface, at the level of our DNA, we are nearly identical.**
> —Spencer Wells, Explorer-in-Residence at the National Geographic Society, Cornell University professor, author of *The Journey of Man* and *Deep Ancestry*, and leader of the Genographic Project

Although we are prone to exaggerate and misrepresent the significance of a few superficial physical differences and habitually break from one another for cultural reasons, modern humans are in fact a young and very closely related species. This is what science says about who we are. Humankind is a family. Dysfunctional, perhaps, but a family nonetheless. According to the US National Genome Research Institute, all people today are *99.9 percent genetically identical*.[166] According to genetic evidence, every person alive today is of African descent.[167] How do these facts align with endless wars and crimes of exploitation and neglect committed in the name of human differences? Genetically, only about one-tenth of one percent separates us as individuals. The scientific process revealed this important and fundamental truth about ourselves, and yet we continue to fear, hate, abuse, and kill our sisters and brothers as though they are strangers from another distant world.

The late Ray Bradbury, author of *Fahrenheit 451* and other classics, described modern humans as the "in-between generation". We are, he believed, "too soon from the cave, too far from the stars". [168] I feel frustration in those words. We have come far and have much to be proud of. But it is also apparent that we are falling well short of our best. Look around at your planet and you will observe fellow humans fighting against themselves and laboring against avoidable contrived problems rather than against real challenges from external sources. Perhaps it is only because so few know who we are that so many are able to believe in the existence overwhelming and insurmountable differences between us. Perhaps if a majority became aware of humankind's real kinships, similarities, common origin, and common dependence on the natural world, we might find ways to live that are more fitting for a highly intelligent species with so few natural barriers to peace and progress. Maybe if more people recognized the profound role of cultures in propagating the irrational beliefs, destructive traditions, and other artifices that do us great harm, we would be less inclined to turn against ourselves again and again for short-sighted or senseless reasons. Knowing ourselves reveals, with undeniable clarity, that we are tied to one another and to all of nature. This awareness offers us a chance to find the maturity and wisdom we will need to find the best ways forward into a challenging and uncertain future. It may be nothing less than the key to our survival.

Chapter 4

WHAT IS EVERYTHING MADE OF?

To attempt answering the question of this chapter, simple and plain as that question may be, we are obligated to enter a strange realm, the eerie sub-universe of the atom where little is as expected and much seems impossible. This is a strange place where something can be in two places at once. Two things can be separated by a vast distance, with no apparent connection, and yet something done to one will cause the other to react in the same exact moment. And, somehow, we can do things in the future that change the past. This chapter, more than any other shows that reality does not conform to human comforts or intuitions—not even close. The very nature of reality, and everything we think we know about it, seems to go out the window the closer we look at atoms. Quantum Mechanics or Quantum Theory, the science of atoms and subatomic particles, shows us the power of science. With the scientific method, by maintaining our allegiance to evidence and observation rather than beliefs and assumptions, we can know, perhaps, anything and everything given enough time, no matter how hidden, unexpected, or mysterious.

The general concept of the atom goes far back in history, at least to the fifth century BCE when Greek philosophers Leucippus and Democritus suggested that all matter was made up of tiny, irreducible or "uncuttable" units called atoms. But while atoms may be the smallest basic unit of matter, they can be broken apart to reveal their different components, of course. Physicists have discovered many distinct parts of atoms. These are called subatomic particles. Of these, the most basic three are protons, neutrons, and electrons. At the center of an atom is the nucleus made up of protons and neutrons. Neutrons have no electrical charge, so they are neutral. Protons have a positive relative charge and electrons have a negative relative charge. The number of protons that an atom has determines its atomic number. Remember the Periodic Table of Elements from middle school? It is arranged based on the atomic numbers of all the elements (different kinds of atoms) in the universe that are known. All atoms are roughly the same size because even those with many electrons have a corresponding number of protons in the nucleus. This means the negative and positive charges are balanced so that the electrons are held in place. Let's be honest, though, atoms are so incredibly small that they are beyond the ability of most people to comprehend. For example, *five trillion atoms* can fit comfortably in a space no larger than the period at the end of this sentence.[169] A single strand of human hair is an estimated *half-a-million atoms wide*.[170] And yet look around. Everything you see is made of these tiny bits of matter, including your body. Atoms make the molecules that make the cells, that make you.

A NEW ATOMIC YOU EVERY YEAR

Your body is a vast and complex transit point for atoms. Always coming and going, trillions of atoms maintain the essential, recognizable pattern that is you, but *each year 98 percent of your atoms are replaced* with new atoms.[171] This exchange occurs via eating, drinking, breathing, sweating, excreting waste, and other bodily functions. It was confirmed in the early 1950s when scientists fed radioactive atoms to experiment subjects and then traced the movement of the atoms. They found that that the new atoms moved to all parts of the body where they were swapped for older atoms.[172]

Almost all atoms in a human body are hydrogen, carbon, nitrogen, and oxygen. While these atoms may be young or new to you, they certainly are *not young or new to the universe*. All the hydrogen atoms in you at this moment were created during the Big Bang—13.8 billion years ago. The other atoms were made in stars, long ago and far away.

Readers might recall seeing the popular illustration of an atom that shows electrons orbiting the nucleus like planets around a star. This is highly misleading and should be purged from everyone's imagination. Electrons do not circle the around the center of an atom in anything like planetary orbit patterns as those kinds of illustrations suggest. The simplest way I can describe it is that electrons exist in an energy field, a shell, or cloud that is far outside of the nucleus. The "Uncertainty Principle" holds that it is impossible to determine an electron's momentum and exact position at the same time. In other words, if we can locate an electron, we can't be sure exactly how fast it is traveling; and if we know its speed, we can't specifically pinpoint its location. Electrons can simply pop in and out of existence.

One of the weirdest things about atoms is that they never really "touch". If matter were to touch in the way we commonly imagine solid objects coming into contact, it would be through the touching of electrons, the outermost part of atoms. But this doesn't happen. Nothing solid, no matter of any kind, every really touches. If you are sitting right now, your rear-end is actually floating above the chair. If you are standing, your feet or shoes are in reality hovering above the ground by the slimmest of spaces. Strange but true. Electrons either attract or repel based on their charge. They don't come into physical contact with one another. So when you "touch" a wall with your finger, the atoms in your finger do not really touch the atoms of the wall. The reason your finger doesn't keep going through the wall, is that the electromagnetic force of the electrons in the atoms that make up the surface of the wall repel or push back against the electromagnetic force of the electrons in the atoms in your finger. All "touching" is really the interaction of energy fields. We feel as it we physically touch things because that is the way our brains have evolved to process it. So this standard way we have of perceiving our interactions with matter around us, something we rely on to function, may be a grand illusion.[173]

Quantum tunneling is another curious aspect of the atomic realm. Electrons can appear on the other side of what should be an impenetrable barrier. We know that this a real phenomenon because of the Sun. Yes, the Sun is conveniently bright and hot because hydrogen nuclei somehow tunnel through an "impenetrable" wall created by the electromagnetic repulsion of their protons.[174] They shouldn't be able to do that, but they certainly do. And, keep in mind, when particles do this tunneling there is not time lapse. They are one side of a barrier one moment, then on the other side. It's not about them being fast but about not really having a precise location.

They are here, there, maybe anywhere. Here's something weirder still, merely observing subatomic particles can sometimes change them. Attempting to measure, or just look at them, can change their state and behavior. Some physicists have suggested that the universe only becomes real when we observe it.[175]

The famous double-slit experiment, first performed by American physicists Clinton Davisson and Lester Germer in the 1920s repeated by many others since, revealed one of the most unexpected discoveries in the history of science.[176] When everyday objects, like grapes or marbles, are shot through a slit at a wall a predictable pattern is created from the impacts. When energy waves are sent through two slits they collide and create a predictable interference pattern on the back wall. If there are two slits through which the objects or waves must pass then the objects or waves will travel through one or the other slit and still create a predictable pattern against the backdrop. But all this commonsense, expected behavior vanishes at the quantum level.

When researchers fire a single electron (an object) at two slits it somehow changes from a particle to a wave, goes through both slits, interferes with itself, and creates an interference pattern on the wall as a wave would. Understandably confused by this, scientists set up detectors that allow them to know which slit the electrons go through. But the act of being observed—brace yourself, here it comes—changes the behavior of the electrons. They go back to behaving like objects and not waves. When observed, matter behaves as we might expect: the electron goes through just one slit, not both. When unobserved, however, particles get weird and act like waves.[177] Physicists still do not understand why or how this happens. How does simply being observed change the behavior of a particle? Does it somehow know that it's being observed?

In 2015 researchers at the Australian National University (ANU) conducted a variation of the double-slit experiment using helium atoms that were scattered by light. They got the same strange results. "It proves that measurement is everything. At the quantum level, reality does not exist if you are not looking at it," said Associate Professor Andrew Truscott of the ANU Research School of Physics and Engineering.[178]

Could it be that reality is not the substantial, reliable, and firm footing that most of us assume it is? Do we make new realities by our actions? Are there many universes in addition to ours? My personal favorite oddity to emerge from Quantum Theory is something called Entanglement. This is absolutely nuts, makes no sense, should be impossible, and yet it happens. Entanglement occurs when two subatomic particles become linked together in a way that changing the state of one instantly changes the state of the other one, too, and it happens *simultaneously*. According to the math, this will happen even if they are many light-years apart. But it should not be possible because nothing, not any kind of energy, light, or information of any kind is supposed to be able to exceed the speed of light. Yet somehow, two seemingly separate subatomic particles react together as one, or communicate between themselves, faster than the speed of life. Quantum entanglement has been shown to be true in multiple experiments. In 2017, for example, Chinese scientists reported using a satellite to test quantum entanglement over a record distances. They said they beamed entangled pairs of photons to three different ground stations that were all more than 1,200 kilometers apart and the results were precisely what Quantum Theory promises.[179]

Those who are finding some or all of this to be difficult to accept may consider themselves in good company. Albert Einstein, one of the greatest physicists ever, had big problems with quantum mechanics, too. He struggled to make peace with its profound weirdness because it was just so completely contrary to *the way things ought to be*. The late science writer and mathematics instructor Amir D. Aczel touched on this in his excellent book *Entanglement*: "Entanglement breaks down all our conceptions about the world developed through our usual sensory experience. . . . To understand, or even simply accept, the validity of entanglement and other associated quantum phenomena, we must first admit that our conceptions of reality in the universe are inadequate. Entanglement teaches us that our everyday experience does not equip us with the ability to understand what goes on at the micro-scale, which we do not experience directly."[180]

Strange as it all may be, Quantum Theory has practical implications. The more we understand the foundation of reality, the more we can do. Discoveries in quantum mechanics can have major implications for future technologies, even if they are not fully understood. Chad Orzel, a physicist and popular science writer, addresses this in his book, *How to Teach Quantum Physics to Your Dog*: "Without an understanding of the quantum nature of the electron, it would be impossible to make the semiconductor chips that run our computers. Without an understanding of the quantum nature of light and atoms, it would be impossible to make the lasers we use to send messages over fiber-optic communication lines." He continues:

> Quantum theory's effect on science goes beyond the merely practical—it forces physicists to grapple with issues of philosophy. Quantum physics places limits on what we can know about the universe and the properties of objects in it. Quantum mechanics even changes our understanding of what it means to make a measurement. It requires a complete rethinking of the nature of reality at the most fundamental level. Quantum mechanics describes an utterly bizarre world, where nothing is certain and objects don't have definite properties until you measure them. It's a world where distant objects are connected in strange ways, where there are entire universes with different histories right next to our own, and where "virtual particles" pop in and out of existence in otherwise empty space. Quantum physics may sound like the stuff of fantasy fiction, but it's science. The world described in quantum theory is our world, at a microscopic scale. The strange effects predicted by quantum physics are real, with real consequences and applications.

Another unexpected thing about the subatomic universe is that it's dominated by *nothingness*. The way that a single atom is structured is nothing short of bizarre. The electrons, wherever they are exactly, are always relatively *very far* from the nucleus. So far, in fact, that every atom is mostly empty space. Pick up the heaviest rock you can find, hold a bar of solid steel in your hand, lean against a brick wall. All empty space, by far. If an atom's nucleus were about the size of a baseball, the outermost reach of the electron cloud of the atom would be about three miles or five kilometers away. Imagine placing the baseball-sized atomic nucleus on the ground and then walking three miles out to the extreme edge of the atom. You would see nothing during the journey. Atoms are 99.9999999999999 percent empty space.[181] But, as small as the nucleus is compared to the overall atom, it contains almost all of its total mass. In a typical atom, the nucleus is *one-trillionth* of the volume but *99.9 percent* of its mass.[182] Having so much nothing inside of every atom means, of course, that *all matter is mostly nothing.* The entire human population, for example—more than seven billion people—could fit into a space with a volume no greater than that of a single sugar cube if all the empty space in their atoms were squeezed away.[183] What, then, is the actual matter, the physical sum, of a single person? Virtually nothing. Even the universe, with all its countless stars, planets, and moons, is mostly a spectacle of empty space.

We may have eyes, ears, and fingertips but we sense only the tiniest fraction of our surroundings. Our large, complex, and powerful brains struggle to understand how everything in our environment is structured and how all of it functions. When we explore reality at the level of the atom it is never more apparent just how limited we are. This is precisely why science is so invaluable to civilization for its practical contributions. Science enables us to see further than we otherwise can and perceive more than could ever hope to otherwise. It also offers everyone of us an endless source of wonder and excitement.

Chapter 5

WHAT IS LIFE?

This is an extremely difficult question to answer well, because it is far more complex than it would seem. There is a good reason that there has never been a universally agreed upon definition of life. The harder one tries to craft a logical, comprehensive, and consistent definition of life, the more elusive it becomes. In the past scientists have presented what were essentially lists of standards to be met for something to qualify as life. Of course, other scientists would then immediately point out the inconsistencies. Is something alive if it is highly organized, consumes energy and grows? Crystals are highly organized. Fire consumes energy and grows. But no one thinks of them as being alive. What about tardigrades (tiny multi-cellular, eight-limbed animals) and some species of bacteria that can go into dormant states for millions of years and do none of the things we associate with life—moving, growing, eating, excreting, reproducing, metabolizing—only to "live" again later?[184] Can something be both life and non-life in the same lifetime?

One of the most obvious and noteworthy features of Earth life is that it has been so successful to date. While specific kinds of life seem to come and go like the wind and biodiversity is declining presently, life in general has proved stunningly resilient and creative, having evoled to thrive many different habitats by many different means. The most comprehensive study ever done on the total volume of Earth life concluded that there are likely more than one trillion species alive right now and only about one-thousandth of one percent of them have been identified.[185] Attach that one-trillion living species figure to the widely accepted estimate that about 99 percent of all species that have ever lived have gone extinct.[186] The mind boggles at how many unique kinds of lifeforms must have inhabited this one planet over the last few billion years. Imagine a comprehensive catalog of all Earth life with 100 trillion entries!

I have been fortunate to experience a tiny fraction of this present Earth life at its most glorious, at least by my subjective human standards. During a night dive in the Caribbean I floated motionless at a depth of 130 feet before a vertical wall of life. To the rear was the ominous blackness of the depths; in front, a thriving and ecologically invaluable coral community. *Coral reef ecosystems shelter about a fourth to one-third of all marine life.*[187] *All twenty-nine of the globally significant coral reefs on UNESCO's World Heritage List are likely to be dead by the end of this century.*[188] As I moved the narrow beam of my flashlight across the face of it, I became an astronaut in zero-g, making first contact on some faraway exoplanet. But I quickly returned because none of this life was alien. Even down there, breathing packaged air in the dark, I was home, with neighbors all around. It was a transcendent moment that moved well beyond simple concepts of pretty colors and physical beauty. I knew it all, even the creatures I could not identify. Life knows life, I suppose.

What struck me most about life in South America's Amazon rainforest is how restless and persistent it seems. It feels as if it is everywhere always. I imagined closing my eyes and then opening them a moment later to find myself covered by vines—but not in a bad way. The weight of green matter is crushing at times—but not in a bad way. I was far from the noise and congestion of civilization but found no silence or space in the Amazon. Life is noisy there, with birds and monkeys filling the air with sounds by day and insects working the night shift. Open space was scarce as plant and insect life seemed to seek out and exploit every possible opening and opportunity. It all might have been scary if it were not so beautiful. *The Amazon rainforest is estimated to contains one of every ten currently known species on Earth. This includes 40,000 plant species, 3,000 freshwater fish species, and more than 370 types of reptiles. More than 2,000 species of plants and vertebrates, have been identified just since 1999.*[189] I struggle to believe how any person could come here and see so much green and not be moved to revere the natural world. *Human activity is rapidly destroying the Amazon rainforest.*[190] I will never doubt the intrinsic value of Earth life because I once walked and slept in the Amazon.

In Australia I explored the oldest living rainforest on Earth. Life has grown, flourished, and towered in the Daintree Rainforest for some 180 million continuous years. No surprise, then, that plants and time were on my mind. *There were nearly 400,000 known land plant species in 2017.*[191] *Approximately 2,000 new species are discovered each year.*[192] *The oldest plant fossils discovered so far are 1.6 billion years old. They were found in central India.*[193] I knelt and looked at decaying foliage on the ground. A stream of ants scurried by, too mission-focused to care about my presence. *Ants have been around for more than a hundred million years.*[194] The greenest ferns I had ever seen were all around me. I have always loved ferns, something about their structure feels old and a little alien to me with their coiled fronds and spores. *Ferns predate trees. They have been around for at least 300 million years.*[195] I touched the trunk of a gigantic tree and try to remember everything I have ever read or heard about trees. They provide habits for numerous lifeforms, cool the Earth, filter the air, reduce erosion, and, for my species, offer beauty and psychological comfort. *Currently there are 60,065 species of trees known to science, according to the first global survey published in 2017.*[196] *Thousands of tree species currently are threatened with extinction due to human activity.* I looked down at the ground around the trunk and thought about the immense underground fungal networks trees worldwide utilize for nourishment, information, and communication. *Approximately 90% of plants and trees share a critical symbiotic relationship with fungi. Vast mycelium networks lay beneath virtually every forest.*[197] But, much as I love trees, it was time that dominated my thoughts in the Daintree Rainforest. It was already more than a hundred million years old when that dinosaur-killing asteroid struck the other side of the planet. To this place, it was only yesterday when *Homo erectus* walked the Earth *Life has been around for so long, four billion years and counting.*[198]

In Africa I saw wild rhinos, giraffes, lions, cheetahs, leopards, cape buffalos, warthogs, and baboons. *Approximately 60% of the world's largest carnivores and herbivores are threatened with extinction.*[199] But it was another human who made me think deeper about Earth life. Somewhere in southern Kenya, my lungs struggle to make use of the few stray oxygen molecules somewhere within the thick smoke I just inhaled. I'm inside a small hut made from mud, sticks, and cow dung. The air is filled with both the bug-repelling fumes of smoldering wood and the incessant chattering of my host, a very old Massai woman. I have no idea what she is saying, but she seems to find me interesting. I find her to be fascinating as well. Could two individuals be any farther apart on the spectrum of humanity? She is an African who depends on the milk and blood of domesticated cows, wild plants, and wild animals for survival. She lives in of a house made of animal poo and apparently breathes smoke on a regular basis. I am an American who depends on grocery stores and restaurants for survival. I live inside a concrete box and breathe a complex and mysterious mixture of industrial chemicals on a regular basis. But before I get very far noting our differences, I hit on something we have in common. We made it. Against incredible odds, both of us exist in this place and time. Every one of our human and pre-human ancestors lived long enough to produce at least one fertile offspring. I look at her through the haze—still yakking on about something—and smile. She and I are winning lottery tickets in the flesh. We are members of a winning team, champions of evolution along with all other species alive this moment. And we share not only the amazing story of human evolution, but the long evolutionary trail of all Earth life, too. We are connected by our common history and nearly identical DNA, not just to one another, but to *all other life*. Everywhere around us we will find only relatives among the animals, plants, fungi, and microbes of this planet. It's a stirring and exciting thought: this woman and I have our humanity in common, but also our connection to life itself—all life. A beautiful thing about possessing at least a bare minimum of geological and biological awareness is that you are never away from home and never far from family.

IN PURSUIT OF ABIOGENESIS

The critical question of *how* life began brings us back to the problem of identifying life. How and when did something not alive become something alive? Was there a dramatic Frankenstein-like-event, during which lightning struck the right molecules in the primordial pool and up sprang something everyone today would agree was life? Did the inert and inanimate become life in an instant, one precise and glorious moment of metabolic awakening? Probably not. More likely there was a long transition between matter of one kind that behaves in certain ways to matter of another kind that behaves in other ways. A likely scenario favored by many experts is that when the Earth was very young and far more chaotic, the chemical processes of some molecules were not life but began to do things that were *lifelike*, such as self-replicate and become increasingly complex over millions of years.

WHAT IS LIFE? There is no universal agreement on the definition of life, and there may never be given the vague-to-nonexistent boundaries between the living and nonliving. *Photo by the author*

At some point during these processes, probably being played out in many different ways in many different places on Earth, something like a cell emerged, thereby launching the long and successful run of bacteria, a reign which continues to this day. A billion or two years later, some bacteria began converting the Sun's energy into *their* energy. Now, with photosynthesis on the scene, life began pumping oxygen into the atmosphere which would then help make the Earth habitable for very different kinds of lifeforms much later, including humans.

Scientists believe one of the key events in the history of life on Earth happened at least 2.7 billion years ago when some bacteria and some Archaea (both are microscopic single-celled creatures) came together and formed a symbiotic relationship.[200] Imagine an archaeon eating, engulfing, capturing—whatever—a bacterium. But the archaeon doesn't simply digest and excrete the bacterium later. Instead the two life-forms fuse and begin working together to become a more complex and capable organism. This was a hijacking, or moment of teamwork, depending on how you wish to view it, with historic consequences. This ancient biological collaboration led to the evolution of eukaryotes, the domain that would one day include plants, fungi, protozoans, animals, and, of course, humans.

Science may never be able to answer how life began on Earth. We can't even say for sure that it even began on Earth. Life on our planet might have come from somewhere else originally and then evolved here. Panspermia, also known as exogenesis, is a real possibility. The earliest life forms could have arrived aboard a meteor or space dust rather than emerging independently here. It's theoretically possible because some microbial Earth life can go dormant for extraordinarily long periods of time in inhospitable environments, like space, perhaps, and then reanimate in a favorable environment. Scientists have revived a 250 million-year-old dormant bacterium in the lab that then successfully reproduced.[201] Some life forms are just *tough*. When the space shuttle *Columbia* broke up on reentry in 2003, killing the crew, a live group of nematodes (roundworms) that were aboard for experimental purposes survived the fall to Earth from many miles up.[202] In 2007 the European Space Agency's FOTON-M3 mission carried tardigrades and exposed them to the cruel, life-destroying conditions of space for ten days. Many of them survived and some even reproduced successfully back on Earth.[203]

The question of abiogenesis, or how life began, is one of the great challenges in science and could well be unanswerable because it happened so long ago, and direct evidence is probably nonexistent. But it is a question so important and so fascinating that we must resign ourselves to pursue it. Whatever took place, happened so long ago that the best we will ever have might be reasoned, logical speculations based on contemporary observations and experiments. It is frustrating, but science currently can tell us more about the origin of the entire universe, nearly 14 billion years ago, than it can the origin of life on Earth that occurred perhaps four billion years ago.

NASA has a deep interest in defining life because astrobiologists at the space administration are looking for it beyond the Earth. One has to know what one is looking for, right? NASA's current working definition of life is short and to the point: "Life is a self-sustaining chemical system capable of Darwinian evolution."[204] Wisely, they didn't make it long and dense, and therefore loaded with inconsistencies. But it's still not perfect. As a habit, I always invoke viruses as the obligatory wrinkle in every discussion about life versus nonlife. Viruses are fascinating and extremely important—somethings. But they meet almost no one's criteria for life because they can't reproduce alone or with other viruses and they lack independent metabolic activity. Yet, viruses are not quite rocks, are they? Viruses evolved for billions of years alongside plant and animal life, humans included, and have had a profound impact on Earth's biodiversity.[205] The evidence of this can be found in the DNA of many life forms today. The human genome is about five to eight percent viral genes.[206] So while viruses may not be capable of fully independent reproduction and evolution, they certainly do evolve and have a big say in Earth evolution. The more one looks, the more blurred this presumed line between life and nonlife becomes.[207]

Ferris Jabr, a contributing writer for *Scientific American*, wrote an elegant and enlightening essay on the challenge of defining life. In it he pointed out the following problems with NASA's inclusion of the requirement for Darwinian evolution: ". . . NASA's working definition of life is not able to accommodate the ambiguity of viruses better than any other proposed definition." He continues:

> A parasitic worm living inside a person's intestines—widely regarded as a detestable but very real form of life—has all the genetic informa-

tion it needs to reproduce, but it would never be able to do so without cells and molecules in the human gut from which it steals the energy it needs to survive. Likewise, a virus has all the genetic information required to replicate itself, but does not have all the requisite cellular machinery. Claiming that the worm's situation is categorically different from that of the virus is a tenuous argument. Both the worm and virus reproduce and evolve only "in the context" of their hosts. In fact, the virus is a much more efficient reproducer than the worm. Whereas the virus gets right down to business and needs only a few proteins inside a cell's nucleus to initiate replication on a massive scale, the parasitic worm's reproduction requires use of an entire organ in another animal and will be successful only if the worm survives long enough to feed, grow and lay eggs. So if we use NASA's working definition to banish viruses from the realm of life, we must further exclude all manner of much larger parasites including worms, fungi and plants.[208]

HOW OLD IS LIFE?

Life is extremely old. There is an unbroken chain of life that extends back *billions of years*, almost to the beginning of Earth itself. By human standards, life has been around for an extraordinarily long, difficult-to-comprehend span of time. *Homo sapiens* is one of the latest arrivals, having evolved only some 200,000 to 300,000 years ago. Life of one kind or another, however, was on Earth at least around four billion years ago. The current record holder for oldest evidence of life are finds in Canada that date back to 3.95 billion years ago.[209] For life to have arisen so early in Earth's history is remarkable, and it's an indication that life might not be so difficult to get going, at least on terrestrial or rocky planets. Maybe it is a common occurrence across the galaxy and universe.

As old as we know life on Earth to be, however, life in the universe could be even older, much older. If it exists, life elsewhere might predate the Earth by billions of years. This prompts speculation regarding from where the evolutionary path of a life form that goes back ten or fifteen billion years old might have emerged. Many experts, cite the high number of galaxies, stars and planets—opportunities for life to begin and evolve—and conclude that it is more likely than not that life is a common feature of the universe. Of course, anything is possible here because to date no conclusive evidence of non-Earth life has been discovered. Therefore, Earth could be one of many planets teeming with life or it could be alone and unique as the strangest planet in all the universe. That seems unlikely in a universe so large, but it's possible. Earth life could also be the first life in a universe that will in the future have life on many other worlds, just not yet. Someplace had to be first, I suppose. Perhaps the most troubling possibility of all is that Earth could be the *last* planet in the cosmos to host life. If that were true, however unlikely, imagine the crime against existence it would be for us to continually squander so much of our intellectual and creative potential.

To move closer toward knowing life and its origin, we may need to abandon the way we tend to think of death. It may seem to us to be a final and permanent state, but this is not necessarily the case for many life-forms. In their book *A New History of Life*, leading biologists Peter Ward and Joe Kirschvink point to advances in science that show just how fuzzy the line between life and death can be. "It turns out there is a vast new place to be explored between our traditional understanding of what is alive and what is not. And this newly discovered region has important implications about the first chapter in the history of life on Earth, telling us whether 'dead' chemicals, when correctly combined and energized, could become alive. Life, simple life at least, is not always alive. But now science seeks to find if there is a place in between. It could be that the first life on Earth came from the place we call death, or from someplace closer to being alive."[210]

WHAT BLEW UP DURING THE CAMBRIAN EXPLOSION?

One of the most important time events in the history of Earth life occurred around 540 million years ago. The animals showed up. And they came in force. The Cambrian Explosion marked the first known mass appearance of complex organisms. It is important to get the general time frame straight when thinking about life on Earth because when one looks around today and sees birds, dogs, cats, cows, deer, elephants, fish, etc. it might be easy to imagine that animals have almost always been key components of the planet's biodiversity. But this is not true.

RISE OF THE ANIMALS. Trilobites are the most the iconic of the early marine animals that lived during the Cambrian Explosion. More than 20,000 species of these armored and segmented creatures roamed prehistoric seas for nearly 300 million years.[211] *Photo by the author*

Animals were late to the party. For more than three billion years, *all* life on Earth had been simple, mostly one-celled life forms. The evolution of multi-cellular, highly complex organisms—animals—was one of the most significant events ever and it happened relatively late in the game. After thriving in the ocean environment, new animals invaded the land some 100 million years later and found success there too.

Some scientists question if there even was an "explosion" at all. Perhaps what we see in the rich fossil record of the time is a bias of evidence. Conditions might have been great for fossilization in a time and place that presents a picture for us today of complex animal forms "suddenly" appearing. Scientists have discovered some life forms that were complex and clearly qualify as animals in the period 100 million years or so prior to the Cambrian period.[212] Trilobites, so prominent in the Cambrian Explosion, could be overrepresented in the fossil record because their hard exoskeletons would have made them much better candidates for fossilization than the soft-bodied creatures that coexisted with them.

Even if complex animals had already been evolving earlier, it does seem that the pace of evolutionary change and the forming of new species quickened dramatically about half a billion years ago. But why? One of the most common possible explanations is that it was a simple arms race. Maybe competition became so intense between the predator and prey of simple life forms that evolution placed a higher-than-ever-before premium on increasingly complex body plans and abilities. Maybe being a blind single-celled creature just wouldn't cut it in some of the rougher ocean neighborhoods. Environmental conditions might have been behind the Cambrian Explosion. Atmospheric oxygen levels increased up to about 13 percent (compared with 21 percent today) at the beginning of the Cambrian Explosion and fluctuated throughout. This increase might have provided a rich energy source for life forms, especially predators who could succeed at catching prey at a higher rate through greater speed, power, and/or endurance.[213] Carbon dioxide levels were hundreds of times higher at this time than they are today. There would have been an extreme greenhouse effect, with higher temperatures than at any other time during animal life on Earth.[214]

A large part of the problem with the quest to define life is probably rooted in bias. We see ourselves as not only different and separate from rocks, water, and stars, but also infinitely superior. Life is better. After all, who would trade places with a pebble? Our perspective is skewed, however. Because we eat, drink, breath, move around, reproduce, and die we naturally would like to assume that those things are the parameters of life. But centuries of scientific work have shown us otherwise. Life is not as clear-cut and obvious as was once thought. The chemical processes of nonlife can be complex and impressive, too. Scientists have created inorganic particles in a lab, for example, that behave a lot like what most people think of as life. These particles "feed" off of light and chemicals, move around, break up, and reassemble.[215] All life is chemistry and physics. It's stuff doing stuff. But on a deeper level the same can be said of nonlife. Look closely enough, and amazing things are going on everywhere and with everything. Atoms never sleep.

All living things are arrangements of atoms—and non-living things are arrangements of atoms. We are talking about *variations* of matter. Science writer Jabr, believes that life is a concept people invented.[216] I think that's about right. This is not to say that living things are somehow insignificant or that the experience of being a human is no better than that of a chunk of iron. Given the choice, I pick human over granite boulder. I have struggled with defining "life" and the best I can come up with is this: Life is what *some* matter does *sometimes*. Or this: Life is matter that behaves in ways that *humans decide* are significant. Because people are alive we elevate the importance of living. Stars may not be alive, but they certainly are important. I don't feel the need to emotionally invest in some dramatic divide between the organic and inorganic. Such a border, if one can be drawn, will certainly blur as robotics and artificial intelligence work continues to progress. It now seems inevitable that humans in the near future will be confronted with machines that look, sound, feel, and act alive. They likely will think in ways that make them seem alive, too. Will they qualify as life? Why not? They will be a collection of molecules, like we are, and they might have something like consciousness. Who are we to deny them living status? Remember that the atoms that ultimately create and recreate you and me anew to some degree each day are inorganic matter (nonlife). These atoms were inorganic matter for billions of years before we were born—and will be for billions of years more after we die. At some time or time during their existence they may contribute to the construction of a lifeform.

Some kinds of atoms (nonlife) might join to become some kinds of molecules (still not life) that might join together to become a cell (life). It all comes down to scaling and timing. It is difficult or impossible to identify a precise time or location where life begins. Think of it this way, if I had all the parts in my garage necessary to make a bicycle, and then began assembling them, at what point would there be a "bicycle" in my garage? Would it be when one or both tires are on the frame? When the handle bars are attached? When every part is in its proper place? Or was there always a bicycle in my garage? I suspect there is no consistent or perfect answer to this. For some it's may be the tires on the frame that do it. For others the attachment of both tires and the handlebars mark the moment of transition to official bicycle status. Some people might argue that it only becomes a bicycle when it becomes rideable and capable of functioning as a bicycle. What do you think? When is a bike a bike? And when is matter alive?

What is commonly thought of by people as life is really just one of countless expressions of matter, ever more unique arrangements of atoms, in a universe with so many possibilities and opportunities. This is not to deflate or degrade the wonder, joy, and beauty that we get to experience sometimes. To the contrary, being more realistic about life, and deciding not to imagine that there is some insurmountable canyon of meaning between the living and nonliving, just might bring our universe into sharper focus.

Astrophysicist Brian Koberlein points out that the deep connections every person shares with the inorganic matter of the universe does not take away meaning from their lives. It could, perhaps, add meaning. "How deeply and subtly connected you are to the cosmos," he says. "It's often said that we are made of stardust, but that just barely scratches the surface. Our bodies are made of the most common elements. Parts of us are as old as the big bang. The DNA in our cells traces a lineage that echoes a billion years of history. Everything about us is an elegant dance of forces and matter, energy and light. Evolution has granted us the ability to look at the stars not just with wonder, but with understanding. Every child born into this world is our species saying live. Survive, for one more generation. Here's a chance to do better than your ancestors. If humanity were scattered across the galaxy, you would only find a single person for every fifty star systems. We are so extraordinarily wondrous and rare. But because we all crowded upon a single world, it's difficult for us to appreciate that."[217]

LIFE CHANGES

One of the fundamental aspects of life is that it evolves. Life changes. Evolution is so important to what life is and how it works that it must be understood if one is to have any chance of understanding life. In the 1970s the late evolutionary biologist Theodosius Dobzhansky famously wrote: "Nothing in biology makes sense except in the light of evolution".[218] Attempting to know life without evolution makes about as much sense as trying to understand automobiles while ignoring everything to do with roads, traffic laws, and the act of driving.

The genetic and fossil evidence is mountainous and clear. The case for life having evolved over billions of years from simple, microbial life forms to the stunning biodiversity we see today is overwhelming. This is not quantum theory. Anyone can grasp evolution with a bit of mental effort. No dramatic suspension of normal perceptions or suppression of common sense is required. A convergence evidence from many sources points to the same conclusion: Life changes over time. Populations change nonrandomly over generations via the unintelligent and indifferent forces of evolution.

MANY PATHS TO SURVIVAL. Over some four billion years, nonrandom but indifferent and unintelligent evolution have produced a staggering wealth of biodiversity on Earth. *Photo by the author*

Evolution is a complex theory and it is neither complete nor perfect—no theory ever is—but anyone who takes some time to learn about it can see that it makes sense. The critical component within this change, the thing that makes evolution so productive and creative, is that there is *descent with modification,* heritable change in populations across generations. Evolution happens when certain traits prove useful enough in the present environment that they may be favored, or *selected*, and passed on to offspring at a higher rate due to their value. In this way, a trait might become common to the species. This is how natural selection changes the look, abilities, and behaviors of a species and creates new species over time. Evolution is about environmental conditions changing the genetic makeup of a population. *Homo sapiens* exists in present form because some populations of some prehistoric hominins lived in total or just-enough isolation from other hominin populations to evolve along their own unique path. One path out of many lead to modern humans today.

In his exceptional book, *The Fact of Evolution,* Cameron M. Smith, a Portland State University anthropologist describes "evolution" as the word we use to characterize the unintended consequences of three independent facts of the natural world:

1. the fact that life-forms have offspring (replication),
2. the fact that offspring are not identical (variation), and
3. the fact that some offspring pass more of their genes on to the next generation than do other offspring (selection).

"Evolution is not a thing," Smith writes, "it is the *consequence* of these three observable facts of *replication, variation,* and *selection.* And because evolution is the logical consequence of these three facts, it is no longer debatable; evolution doesn't just happen, it *has to* happen."[219]

Inevitable reality it may be, but large numbers of people do not accept that life evolves. According to a 2017 Gallup study, 38 percent of US adults (more than 90 million people) say that humans are only about 10,000 years old and have only ever existed in current form.[220] The evidence, however, makes it clear that anatomically modern humans are 200,000 to 300,000 years old, and there have been many versions of humans, from *Homo habilis* and *Homo heidelbergensis* to the Neanderthals and *Homo floresiensis.* Jack Horner, the prominent paleontologist who made some of the most remarkable dinosaur fossil discoveries ever, believes he and his colleagues share some of the blame for this. "Scientists have done a bad job of teaching people what evolution is," he told me. "I think it's about time for us to figure out how to teach it to people. [Evolution] has to do with similarities. We look at the human being and we look at the ape and we can see that they share more common features with each other than they do with anything else. So, just like with a brother and a sister, we can assume that they have a common ancestor."[221]

The elephant in the room here is religion. Not all, of course, but some versions of some religions encourage or require followers to reject the foundation of modern biological science. Some even see it as more of a moral issue than one of describing reality. Paleoanthropologist Donald Johanson, discoverer of the famous Lucy fossils, suggests that it might help if people tried to separate religion and science and view them as two distinctly different ways of obtaining knowledge. "There are two very different ways to try to explain our human existence," he said. "One of them is the faith-based endeavor and that depends exclusively on how one is raised. If you are raised as a Hopi Indian, then you will learn the myths of creation of the Hopi Indians. If you are raised in a tribe in South America, you will believe that story is the truth. And if you are raised as a Catholic, then you will believe that is the true answer. All of that is based on experience, how you grow up, and how you are taught. It is based on faith. We don't subject religious ideas to the same rigorous investigation as scientific issues. Regarding evolution, we are looking at the scientific evidence for how we came to be who we are today. It is a fact of the natural world that all animals, plants, and insects have gone through a process of evolution by means of natural selection. We don't look at gravity, which is a fact, and ask if it is moral or immoral. It is simply a fact."[222]

FIVE COMMON MISCONCEPTIONS ABOUT EVOLUTION

- "A random process like evolution could not produce the kind of complexity, biodiversity, and beauty that we see in nature." This objection to evolution comes in various forms: In social media circles, I have seen references to monkeys randomly banging on typewriters for a billion years but never producing beautiful poetry and tornados ripping through junkyards but never managing to assemble a jumbo jet. These are old challenges that never worked and still don't work. The conclusion they are supposed to lead one to is that evolution is invalid because it can't randomly throw molecules together to build complex elephants and butterflies, same as the monkeys and tornados can't produce meaningful complexity. The problem with this is that *evolution is not random*. I suspect that some people believe that "blind" and "unintelligent" evolution equates to "random" Evolution. But it does not. The genetic mutations that create variation are an important part of evolution and they may be random, but natural selection is not. Various environmental pressures are *selecting*, over many generations, those traits that boost survival and reproduction rates for a population. Natural selection is not a random process.

It also is important to be aware that complex lifeforms do not just spring into existence. A human born today, for example, is the product of more than his or her parents. That baby was 3.5 billion years in the making. There was an immense stretch of time, many stages along the way, and a tremendous amount of non-random evolution.

- **"If people came from monkey and apes, then why are there still monkeys and apes?** This is repeated often and seems to confuse many people. It can easily be explained, however. The Theory of Evolution does not claim that humans evolved from modern monkeys and apes. Based on very strong fossil and genetic evidence, it seems clear that modern humans, apes, and monkeys share common ancestors long ago, but we have been on our own respective evolutionary paths for millions of years.

- **"Evolution is like climbing a ladder of progress."** Evolution is not synonymous with improvement. There is no foresight, no plan, no goal, and no justifiable ranking of superior and inferior lifeforms in strictly evolutionary terms. Evolution is best thought of a reaction to what happened over previous generations. It is not preparation for the future. It can't be because the future is unknown.

- **"Some species are more evolved than others."** In any given moment we can't declare that some living species are better or *more evolved* than others because everything that is alive now is an evolutionary winner. All current life is on equal terms, in this regard. Extinction is the only valid measure of defeat. A human being with her massive brain, opposable thumbs, and bipedal stance is no better in an evolutionary perspective than beetles or plankton because all three have made it this far. If anything, it's a tie. Environments always change and those species who can't keep up become extinct. For example, should an asteroid hit, supervolcano explode, or something else bad happen next week, flashy *Homo sapiens* could well be wiped out, and the beetles and plankton might be left to run things. For any lifeform, today's impressive traits could well be irrelevant or even harmful when the environment changes—and, sooner or later, it always does.

- **"Intelligent Design makes more sense than evolution."** Intelligent design (ID) is the claim that life is too complex to have been the result of evolution. Because some structures or processes defy satisfactory explanation today, life must have been engineered by some higher intelligence. This is nonsense. Even if Earth life were intelligently designed it still would be wrong to accept the ID claim today because it has no evidence or logic behind it. It would be a lucky guess at best. The intelligent design idea is anti-science philosophy and intellectual surrender.

  Unsolved mysteries and unanswered questions in the biological sciences prove nothing other than the need for scientists to keep working. Many things that are now well understood about matter, life, Earth, and the universe seemed *irreducibly complex* to many people not so long ago. No one could figure out every incremental step and processes that made sense of continental drift. Should scientists have given up then and declared too complex to be anything other than something that was intelligently designed? Today, the answers to many once "impossible-to-answer questions" are casually taught in high school science classes. Ignorance is never an explanation. From the scientific perspective, ignorance is the call to stay it, to continue experimenting and exploring.

Paleoanthropologist Tim White worked on the Lucy fossils with Johanson and has led teams that discovered some of the oldest *Homo sapiens* fossils to date. I asked him if the current mass rejection of evolution in the United States and many other societies disappoints him. "It frightens me more than disappoints," White said. "Understanding evolution, understanding that the biological world that includes us evolved, is essential. This is true in virtually every field of inquiry, fields as disparate as medicine, agriculture, and sociology. Education is the key to improving awareness."[223] I suspect that the current problem some religions have with evolution will not last. One would think that the *origin of life*, rather than the fact that life evolves, would be their primary concern anyway. Most likely, these particular religions will evolve and accept evolution in the same way many eventually have accepted that Earth is not at the center of the solar system or the universe.

ATOMS WITHOUT BORDERS

Recall our exploration of matter and "reality" down at the atomic level in a previous chapter (Chapter 4, "What Is Everything Made Of?"). Even a passing familiarity with quantum mechanics makes it difficult to view life as the be-all and end-all of existence. Once we know something about the origin, age, journeys, dimensions, and behavior of atoms, anything has the power to excite. "Truthfully, that which we call life is impossible without and inseparable from what we regard as inanimate," argues science writer Jabr. "If we could somehow see the underlying reality of our planet—to comprehend its structure on every scale simultaneously, from the microscopic to the macroscopic—we would see the world in innumerable grains of sand, a giant quivering sphere of atoms. Just as one can mold thousands of practically identical grains of sand on a beach into castles, mermaids or whatever one can imagine, the innumerable atoms that make up everything on the planet continually congregate and disassemble themselves, creating a ceaselessly shifting kaleidoscope of matter. Some of those flocks of particles would be what we have named mountains, oceans and clouds; others trees, fish and birds."[224] Yes, with a bit of awareness about what matter is and how it works, a lone rock on the ground becomes an epic story, a wondrous and mysterious package of intoxicating weirdness. Life, however we define it, is endlessly fascinating, no doubt, but so, too, is the rest of existence. I'm not suggesting we take life down from its pedestal or revere it any less. My point is that, in this universe, *everything* is amazing.

THE LIFE WITHIN

While some scientists continue to look farther out into the universe and deeper back in time than ever before, other scientists have been looking closer to home and learning about another fascinating and important universe. This one is much smaller, and it's not filled with stars but with trillions of microscopic creatures. The human gut microbiome, a vast community of diverse microorganisms that live in the intestines, has been called the "forgotten organ" for the neglect researchers have shown it in the past despite its critical organ-like functions.[225] But this is changing fast as our alien world within is finally beginning to get the attention it deserves given its prominent role in human health. It is vital for everyone to know about the gut microbiome and understand that life not only saturates our planet, but our bodies as well. Just as arthropods, reptiles, amphibians, fish, and mammals evolved to survive in different environments so too have thousands of species of microscopic lifeforms evolved to live inside of us. Together, we co-evolved for millions of years and have become so interdependent on one another that today we could not live without them and most of them could not live without us. Learning about the life within us goes a long way toward answering the question *What Is Life?* because it is such a powerful demonstration of the nature of living matter. Life is the story of competition, collaboration, and connection. One lifeform touches another, touches another, ad infinitum.

Best known by scientists today are the bacteria of the gut, representing thousands of distinct species. But they are not alone. Archaea, viruses, and fungi are present, too.[226] It's as if each one of us carries around something like a three- or four-pound rainforest or coral reef in our gut. Pause and let this soak into your neocortex for a moment. *There is a complex living community of diverse creatures thriving inside of you right now.* And you need them to live. The healthier and more vibrant and diverse this ecosystem is, the healthier you are. In 2018 the results of an important study were published that showed the link between gut health and overall health. A US and Chinese team of scientists looked at more than a thousand people revealed a strong correlation between very healthy old people and very healthy gut microbiomes. Older people who were in excellent health consistently had high microbiome diversity.[227] Just as it is with a rainforest or coral reef, more biodiversity generally equates to a healthier and more resilient ecosystem. "This demonstrates that maintaining diversity of your gut as you age is a biomarker of healthy aging, just like low-cholesterol is a biomarker of a healthy circulatory system," explained one of the researchers.[228] Scientists still do not know the direction of cause and effect. They are trying to learn if the gut makes a healthy person or the other way around. I suspect this will turn out to be a two-way street; a case of mutual dependence. As your gut goes, you go.

The trillions of deep residents in your body cooperate and compete among themselves in complex ways that are still mostly unknown.[229] They also help shape your life in many profound and surprising ways. For this reason, we cannot think of them as parasites, tenants, or even welcome guests because they are so much more. When all is well, these creatures are your partners in biology, something like a close family member who has an extremely important role in your life. These lifeforms help modulate your immune system, regulate your body's metabolic activities, influence the quality of sleep, may be influencing how lean or fat you are, and more.[230] They communicate with your brain.[231] Scientists have shown that manipulating the volume/ratio of just two bacterial species in the intestines of rats causes them to demonstrate anxiety-like behavior at higher or lower rates. One remarkable study showed that simply consuming a fermented milk product twice a day for four weeks *changed the brains* of human test subjects. The fermented milk contained four species of living bacteria: *bifidobacterium, streptococcus, lactococcus,* and *lactobacillus.* Somehow, this made the test subjects react more calmly to various face images they were shown while having their brains scanned.[232] Exactly how gut microbes effect brain activity is not known. Regardless of the many remaining mysteries, however, you can be sure that microorganisms are profoundly important because they not only share the journey that is your life, they help guide it.

Babies come into the world with sterile intestines and immediately begin picking up microbial guests that go to work establishing their ecosystem. This occurs during the birth process, through breast feeding, eating, and contact with objects and other people. Typically, by the end of the first year of life a baby's gut microbiome is already unique to the individual and by two or three years it has approximately the same characteristics (type of microorganisms, ratio of species) as it will when the person is an adult. This first year or two period seems to be crucial to long-term health. A disruption of the gut microbiome's early development seems to be a "critical determinant disease expression in later life," according to mounting evidence.[233]

Many people today probably are aware that this complex community is present and essential to the normal daily processing of food. But it also helps destroy disease-causing organisms and produce vitamins.[234] This only scratches the surface, however. In recent years scientists have been finding out that the gut microbiome influences not only physical health but mental health, too, and in ways far beyond what was once thought possible. For example, I found an intriguing study that showed rates of ADHD or autistic spectrum disorders in young teenagers who had been supplemented with probiotics in infancy fell to zero percent.[235] Some gut microbes produce hormones and neurotransmitters (the chemicals nerve cells use to communicate with each other) that are identical to those produced by the human host. These microscopic creatures that live inside of us can influence our memory and even *our thinking*. Scientists have linked the gut microbiome to allergies, multiple sclerosis, depression, anxiety, autism, asthma, Crohn's disease, ulcerative colitis, Parkinson's disease, irritable bowel syndrome, colon cancer, type 1 diabetes, type 2 diabetes, coeliac disease, atherosclerosis, Alzheimer's disease, rheumatoid arthritis, stroke, alcoholism, chronic fatigue syndrome, fibromyalgia, and other disorders.[236] A remarkable experiment conducted on mice demonstrated the power of the gut-brain connection. Researchers fed Lactobacillus rhamnosus, a bacterial species found in some yogurts, to mice and then compared these mice to other mice that were not the given this bacterium. When placed in stressful conditions, the mice who had consumed Lactobacillus rhamnosus showed less panic and anxiety and did not give up when forced to swim as quickly as other mice. The researchers said the effect was very similar to that seen in mice given antidepressant drugs. This was not just about observing and interpreting behavior. The researchers also tested the brain chemistry of the mice and found that the probiotic-dosed mice had only about half the amount of the stress hormone corticosterone in their blood as the other mice did.[237] In the coming years and decades more will come to light and we will understand the connections better but it's accurate to say right now that the gut microbiome is a big part of your life and you can decide through your actions to either help or hurt its health.

Scientists learn more every day about what constitutes a healthy gut microbiome, how it determines aspects of our health and function, and what things we can do to help it help us. Pay attention to new discoveries coming of out this exciting branch of biology and medicine because the gut microbiome directly impacts you every moment. For example, it is now known that the gut microbiome influences how patients respond to cancer immunotherapy.[238] The day seems to be coming fast when cancer therapies, pharmaceutical prescriptions, and other treatments will come *after* a patient's gut microbiome has been analyzed so that everything can be tailored to the patient's unique ecosystem. Serious diseases might be anticipated decades in advance and prevented by a simple reading of a stool sample and then taking appropriate steps. Fecal microbiota transplantation (FMT) is still being studied to determine safety and effectiveness but appears to be very promising. As bizarre and unappealing as it may seem, simply implanting rectally or orally consuming a small bit of fecal matter from a donor with a healthy gut might improve or cure a patient's disease.[239] Researchers are racing to develop more effective drugs that are based on what microbes do in the human gut.[240] There are a number of companies now that will take mail-in fecal samples from you and send back a spreadsheet reflecting the state of your microbial inhabitants. However, some scientists warn that sample collection and testing is a delicate process that can easily become corrupted and that these companies, in general, are attempting to reach too far ahead of the science.[241] Understanding everything about the gut microbiome and positively altering it based on sound data sounds great, of course. The future can't get here quick enough. But what can we do for ourselves right now? What does the best current science say about improving the diversity of our gut microbiome?

Mountains of scientific evidence have made clear that consistent exercise reduces disease risk and promotes overall health. But there seems to be a lot more to this than heartrate and burning calories. Now researchers think that a significant reason why physical activity is so beneficial to people is because the gut microbiome loves it. There is something about inhabiting an active host that encourages and enables gut flora success. Research suggests that exercise has a positive impact on the gut. It seems to significantly drive up overall diversity and elevate the numbers of beneficial bacteria. Remember this: people who exercise on a regular basis tend to have healthier gut microbiomes.[242] This is powerful information. You can walk, run, bike, swim, and lift your way to building a strong and flourishing gut microbiome.

Probiotics products available to the public have exploded in popularity in recent years and may rise above $60 billion in sales by 2023.[243] But before you run to the local health-food store to buy some high-priced mystery concoction from an insufficiently regulated industry, consider giving *food* a chance. The experts are clear on this point. Scientific studies show that the human gut microbiome responds very well to a diet that is rich in vegetables and fruits. The life in your gut benefits when natural, whole foods reach the intestine largely intact and ferment. The typical American or Western diet is so bloated with industrial junk food that is does not happen enough for many people. Too much unrecognizable and indigestible "food" moves by unsatisfied microbes that were looking for a healthy feast. For millions of people, the modern diet includes too much animal protein and fat and too little vegetables and fruits. This diet is associated with low gut microbiome diversity, which is an indicator for poor health and disease risk. Researchers have shown that the typical Western style diet degrades overall diversity, decreases the total number of microbes, and specifically shrinks populations of the beneficial *Bifidobacterium* and *Eubacterium species.*[244]

WHAT DOES SCIENCE SAY ABOUT PROBIOTICS?

Probiotics are live microorganisms meant to be consumed to support or build up the "good" species in the gut. This exciting and attractive potential means of disease prevention, treatment, and health improvement is loaded with potential. Unfortunately, the supplement industry is largely unregulated, and, as a result, is so inundated with fraud and quackery that there is no way to be sure which probiotic products are useful or safe. For example, one study analyzed 14 different commercial probiotic products and only one of them contained the same species of microorganism named on the label.[245]

John Cryan, a principal investigator at the APC Microbiome Institute in Ireland, has done important and widely cited research on the gut microbiome. He is also a coauthor of *The Psychobiotic Revolution: Mood, Food, and the New Science of the Gut-Brain Connection.* I asked him for his take on commercially available probiotics today. "Most probiotic products available are not probiotic," he explained. "A probiotic is a live microorganism that when taken in adequate amounts confers a health benefit. It is this last part that lets many products down as the clinical trials haven't been carried out not to mention experiments to see if they even survive the stomach [to be effective, the bacteria in the probiotic must reach the intestines while still alive]. . . . Diversity in diet is key to having a healthy microbiome, that and fibre. Fibres, especially in vegetables, act as a key prebiotic on which our microbes act to produce essential health-promoting chemicals. Avoid processed food as much as possible and increase the intake of fermented food like Kefir."[246]

Many basic questions have yet to be answered regarding probiotics. Overall, as far as the public is concerned, buying gut-health-boosting drink off the shelf remains more of a hopeful concept than a fact-based reality. For example, researchers have not yet identified and studied all beneficial microorganisms thought to live in the intestines or what specific microbial species ratios are best for optimal physical and mental health.[247] But the potential is obvious and exciting. "We are really just at the beginning of a what we have coined a 'psychobiotic revolution,'" said Cryan, "but are pretty confident that targeting the microbiome will be an essential component in ensuring wellness and building resilience to stress in years to come."

The Mediterranean diet is dominated by plant-based foods and has been linked to highly diverse and healthy gut ecosystems.[248] This makes perfect sense, of course, if one considers who humans are. We are, remember, hunter-gatherers who evolved to eat a diverse diet that is rich in fruits and vegetables. Only very recently have so many people been eating very limited, plant-poor diets. This is one reason why hunter-gathers in the past likely had and hunter-gatherers today do have far more diverse gut microbiomes that modern people.[249] They tended to eat a healthier diet and were closer to nature in daily life which exposed them to more microorganisms. According to one study, consuming commercially available garlic powder may kill or hinder bad gut bacteria and help some of the good gut bacteria.[250] Consuming excessive amounts of sugar, typical in developed countries today, is bad for many reasons, of course. But did you know that it is bad for your gut microbiome? A high-sugar diet has been shown to alter the microbiome of mice in ways that impaired their mental abilities.[251] There is good evidence that fermented foods (such as yogurt, kimchi, kefir, and sauerkraut) seem to be beneficial to the gut microbiome, though it's not clear how much good they do.[252]

Why do we often hear that some foods are great for us because they have a lot of fiber? What's so important about fiber? Why should we care about eating high-fiber foods such as lima beans, lentils, blackberries, blueberries, pears, and apples? Fiber matters for a few reasons but none are more important than what it does for the gut microbiome. Fiber is a type of carbohydrate that does not get broken down well in the small intestine. This means it moves through to the large intestine where trillions of microorganisms put it through the process of fermentation. And their ranks swell as a result.[253] It may unappealing to think about having a compost heap in your abdomen, but it's a good thing. You want it. Commit this to memory: dietary fiber is a good source of valuable "microbiota accessible carbohydrates."[254] Fiber is the key fuel for healthy gut microbiomes. When you eat blueberries or a pear, you are eating something your little friends will love. University of Michigan experts recommend eating about 25 to 30 grams of fiber per day.[255] They also encourage caution, however, for those who have been living on a diet of hamburgers and hotdogs for some time. Don't overwhelm the system in a flurry of healthy eating. Give your microbes a chance to catch up.

Michael Pollan, author of *The Omnivore's Dilemma* and *In Defense of Food*, wrote an essay in the *New York Times* in which he noted that gut microbiome experts tend not to use specialized probiotic products available in stores. They tend to consciously lean on the components of a microbiota-friendly diet that are familiar to us: fruits and vegetables. "Viewed from this perspective," Pollan writes, "the foods in the markets appear in a new light, and I began to see how you might begin to shop and cook with the microbiome in mind, the better to feed the fermentation in our guts. The less a food is processed, the more of it that gets safely through the gastrointestinal tract and into the eager clutches of the microbiota." Pollan continues:

> "This is at once a very old and a very new way of thinking about food: it suggests that all calories are not created equal and that the structure of a food and how it is prepared may matter as much as its nutrient composition. It is a striking idea that one of the keys to good health may turn out to involve managing our internal fermentation. . . . The successful gardener has always known you don't need to master the science of the soil, which is yet another hotbed of microbial fermentation, in order to nourish and nurture it. You just need to know what it likes to eat — basically, organic matter — and how, in a general way, to align your interests with the interests of the microbes and the plants."[256]

Something to *avoid* doing in support of your gut microbiome is to abuse antibiotics. Do not allow them into your body unless absolutely necessary. Properly used, antibiotics are great, of course. They have saved countless lives. But doctors have been overprescribing them for decades. According to the CDC (Centers for Disease Control), one of every three antibiotic prescriptions written by doctors is unnecessary.[257] This is a big problem for the gut microbiome because antibiotics are tremendously destructive to internal microorganisms.[258] They are like bombs that kill indiscriminately, with no regard to species. The good and innocent die along with the bad. One scientist described taking antibiotics as like setting fire to an entire forest to control the weeds.[259] Although the gut microbiome will recover, researchers have found that it may never fully recover to its former state.[260] Keep in mind that antibiotics are anti-*bacterial* and do nothing for viral infections like the flu and common cold infections. If you are sick, think about what you probably have and what you probably need. If necessary, help your doctor make the right recommendation for you—and for your gut microbiome.

There are no excuses, no room for neglect, doubt, or delusion here. You are host to a bustling city of microbes and if they aren't happy and healthy, you probably won't be either. There is may be more ignorance than knowledge about the gut microbiome right now, but, thanks to science, we know enough to take positive action. Eat a mixed diet heavy in vegetables and fruit, exercise consistently, and avoid unnecessary or excessive use of antibiotics when possible. Acknowledge who and what you are. Early in life, every human is colonized by invisible lifeforms and turned into a superorganism. Every woman and every man is a bipedal platform loaded with a magnificent ecosystem of entangled life. We may not be able to look at this community in the mirror every morning (probably a good thing), but never forget that it's there. Let your knowledge and awareness of it provide you with more motivation to do the positive things necessary for your mutual wellbeing.

THE ORGANIC GREAT UNKNOWN

While scientists continue working to learn more about the mostly mysterious life inside of our bodies, it is important that we be aware of how little is known about all the life beyond our bodies, too. Scientific surveys and analysis of existing data indicate that as much as 86 percent of living land species and 91 percent of living ocean species remain undiscovered and unstudied.[261] Ponder this for a moment: Right now, in the twenty-first century, the majority of living species on our planet are a total mystery to us. There is a lot of work to be done.

The challenge is greatest regarding microscopic life. Microbes were the first living creatures on Earth four billion or so years ago and they help sustain life on this planet today in many ways, such as regulating ecosystems and recycling dead organic matter. Microbes define Earth's "habitable zone". Where there is life, any place on Earth, it will include, or exclusively be, microbes. It is impossible to imagine anything close to the natural world existing without bacteria and other microbes performing their constant, universal, and in many ways dominant role.[262] Scientists estimate that there are as many as *one trillion microbial species* alive on Earth right now.[263] To be clear, this means there are some one trillion *distinct kinds* of microscopic creatures, with each species having many trillions of living members. Given such stunning diversity and abundance, it is not so difficult to understand how the smallest life of all has found success in virtually every Earth environment from the upper-crust to the stratosphere. For some perspective, there currently is only one human species with a population approaching eight billion individuals. There was a time in our past, however, when multiple human species shared Earth at the same time. That may seem a bit weird but try to imagine how bizarre it would be if a hundred or a thousand distinct versions of us coexisted today—or one trillion. Microbes are everywhere life can be. They are plentiful, diverse, and we barely know any of them.

Some experts estimate that *98 percent* of living microbial species currently are unknown to science.[264] An estimate like this can be made by analyzing and projecting from samples, dispersion patterns, and discovery rates. This is significant because it represents a staggering ignorance about the most important life forms on Earth. And things are no better closer to home. We know relatively little about the microbes that live *inside of us*. In 2017 Stanford researchers reported that *99 percent* of the DNA fragments they surveyed in human blood samples were from lifeforms that had not yet been identified or studied.[265]

Our personal health and even the survival of humankind are tied to microbes in numerous ways and yet we have met and come to know almost none of them. When we tackle questions about who we are and how life works, it is necessary to give due consideration to the smallest, most abundant, resilient, and creative life of all—the microbes. They thrive *beneath* the deepest seafloor. They ride the high winds of the stratosphere above. Not only are they on us and inside of us, they surround us, too. Each person emits about one million biological particles per hour. This constant flow of organic exhaust creates a unique microbial cloud that surrounds and shadows us every moment of our lives.[266] No human is ever alone. And the fact that we don't know anything about most of the microbial life on Earth, including those inside of us, means we live in semi-darkness. Biological dark matter saturates our existence and learning more about it is one of the key scientific challenge of this century. Being largely ignorant about this extraordinary and key feature of our existence is like drifting in the middle of an ocean with no understanding of the concept of water or any factual knowledge about it. Scientists have work to do and, meanwhile, everyone can benefit from learning as much as possible about what is known to date. Our daily dance with microscopic life impacts our mood, sleep patterns, food digestion, immune system performance, and more. The more we know, the better we might interact with them to enhance our health and happiness.

THE GREAT CONTINUUM

If people were to only remember one thing about life from this chapter, something that might intrigue, excite, or inspire them long-term, it probably would best be this: *Life is connection*. The origin mystery aside, life comes from life, cells from cells. As humans, the most thoughtful, self-reflective, and analytical beings on the planet, we can enhance our lives by making an effort to learn, recognize, and *feel* the links we share with all other life today, all life before us, and the life to come. Everywhere, we are surrounded by ancestors and cousins. I don't know about you, but I feel great and look pretty good for someone who is at least four billion years old. I was once a sensitive Australopithecine, a drifting fungal spore in the stratosphere, a tough sea sponge at the bottom of an ancient sea, and a creative bacterium. In many ways, I still am those things. And so are you. All species within the three domains of life on Earth (Archaea, Bacteria, and Eukarya) share twenty-three universal proteins. Statistically, this is virtually impossible—unless, of course, all life today shares a universal common ancestor.[267] "Self-transcending life never obliterates its past: humans are animals are microbes are chemicals,"[268] declared the late biologist Lynn Margulis and Dorian Sagan in their excellent book, *What is Life?*. "We share more than 98 percent of our DNA with chimpanzees, sweat fluids reminiscent of seawater, and crave sugar that provided ancestors with energy 3,000 million years before the first space station evolved. We carry our past with us."[269]

Life is never really alone on this planet. It shares, steals, competes, and cooperates to form complex ecosystems. It is connected in profound and meaningful ways today and interconnectedness has been a standard feature of life here for billions of years. The building blocks of life connect it to something even grander, the universe itself. Life flows along an uninterrupted river of atoms that, for brief periods, at least, do a thing we call *live*. Knowing something about our personal journey, as well as the travels of all life and all atoms, offers us additional depth to the experience of existence, pleasant fascination, and perhaps even a bit more meaning to our brief moment in the Sun.

Chapter 6

HOW DID WE GET HERE?

I t is one million years before now and you are standing in waist-high grass at the tallest point of an elevated plane somewhere in Africa. You walked up here because it offered the best view of the area. But you see nothing of interest. Nothing moves but the grass. There are no sounds but for the occasional blast of wind. Everything changes, however, when a human form slowly rises from a crouched position no more than twenty meters away.

Fully upright now, the person stands motionless. Her body is lean, muscular, statuesque. It projects obvious power and function-ality. She looks like she could give anyone or anything a respectable fight and then run ten miles home to tell the story to her clan. You notice that the dark skin of her body is in sharp contrast against the light-brown grass and wonder how she could have been so close without you noticing. She rotates an apple-sized stone in her right hand with delicate fingertip touches. She spins it too fast to tell if it's a chiseled stone tool or just some rock she picked up. You esti-mate her height to be about five feet or 1.5 meters tall. She probably weighs around 150 pounds or 70 kilograms.

The woman glances around and takes a few steps forward. There is grace and precision in every movement. Clearly, she is human and yet the humans you have known in your time don't move quite like this. Maybe elite ballet dancers and worldclass hurdlers do, but not many others. Her face is different. It protrudes more than the people you know in your time. Modern humans, you suddenly realize, have very flat faces. Her eyebrow ridges are significantly thicker than you are used to seeing as well, and her forehead is noticeably shallow. Still, overall, she's clearly human, more alike than different.

She freezes in place. Only her eyes move as she scans the wil-derness. With obvious determined efficiency she searches for anything that might translate to danger or opportunity. You are close enough to appreciate the intensity of the effort. This is no dumb, stumbling cavewoman. You can sense her awareness and intelligence. Her brain is on fire now, in full search mode. Its pattern-recognition system strains to pick out the all-important profile of an animal body. She carries in her mind an extensive catalog of shapes and sizes that can be drawn on in an instant to help her determine if an image represents the chance to eat or the possibility of being eaten.

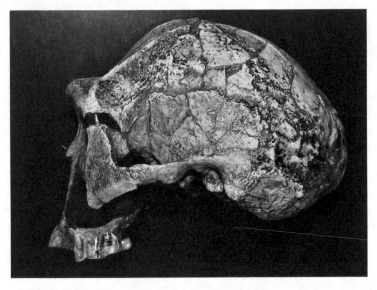

MOVERS AND SHAKERS. With a brain that was about two-thirds the size of modern humans, *Homo erectus* achieved remarkable feats to put our genus on a path of invention, innovation, and exploration. *Photo by the author* of a cast of Khm-Heu 3733, a fossilized *Homo erectus* skull discovered in in Kenya.

She makes a noise. Was it a whisper, a word? The woman is excited. Her eyes fix on something in the distance. She squints and leans forward a bit, presumably zooming in on some hint of an anomaly. More soft sounds escape her lips. But you can't tell if they are words, real language, or just mindless grunts. You look in the direction of her attention and see nothing but the same endless sea of windblown grass and occasional tree. The woman turns around

180 degrees and fires off three restrained, high-pitched shrieks. A similar series of sounds come back soon after. To your surprise, several more people you didn't know were there emerge from the grass a few hundred meters away. Five males and three females, each armed with a sharpened stick and/or stone of some kind, walk toward the woman. Their movements are quick but silent. After an exchange of knowing glances and hand gestures they move down in the direction of whatever it is the woman saw.

Over the next hour you watch these nine humans methodically stalk, surround, and kill a cape buffalo that had become separated from its herd. You marvel at their confidence and the obvious sophistication of communication and coordination. Two of them prowl the perimeter, probably watching out for any of its herd mates that might attempt a rescue. Three others move directly in front of the animal. They beat the grass with sticks and shout. One female waves a leafy branch back and forth like a battle flag. The cape buffalo seems sick or old but still willing and able to fight for its life. It takes a few steps toward the people and snorts in anger. It fails to notice, however, the remaining four people now moving in from behind. Like a team of special ops soldiers from the future, they execute their mission to perfection. As the four in front continue their distractions, the rear four pounce. They sprint the final ten meters, jam sharpened sticks into the animal's neck, and then dash away to safety. With precise synchronization, the people in front intensify the distractions to ensure that the attackers get clear of the animal.

One of the sticks found a major artery and blood spurts from the wound. For the next twenty minutes or so the buffalo trots around in anger and/or fear. The people are careful to keep a safe distance but every time the animal stops they move in just enough to agitate it and keep it stressed and moving. When the animal finally falls, they move in and finish the task. Now, even from your position a couple of hundred meters away, you can tell that they are talking. Words, conversation, something more complex than basic calls are being exchanged. It also is clear they are pleased with their success. Four of them sit around the big animal and butcher it with stone tools while ranting and chattering with one another. A tall male stands up on a boulder, raises both arms above his head, and screams in triumph. You take a small step back. These are amazing people. They seem capable of doing just about anything they put their minds to.

Welcome to your genus. These people are members of *Homo*, the remarkable primate genus that would introduce a new path to survival to the planet, that of brainpower, language, and culture. Specifically, they are *Homo erectus* or *Homo ergaster*, the African version of *Homo erectus*.

Having carved up the animal and shouldered limbs and chunks of meat, the group trots at a moderate pace back up the incline toward your position. They run by in single file, close enough for you to get a good look at each face. You know them. In their eyes is the unmistakable light of human intelligence and emotion. You feel the strong impulse to kick off your shoes and join in, to run with them to wherever it is they are going. But you smile and think better of it. You don't belong here. They are family, however, and you owe them everything. Their existence is forever tied to yours. The struggles and successes of *Homo erectus* both shaped and made your life possible more than a million years later.

Over the next several minutes you watch the foremothers and forefathers of modern humans run toward the far horizon. You admire the way they skim the ground with light, efficient strides. Finally, the tall grass in the distance swallows them up and they are gone. You think about how far they have come and how far they will go. With intelligence, creativity, toughness, courage, teamwork, and a bit of luck, *Homo erectus* will survive. Predators, supervolcanoes, storms, disease, climate change—all that nature throws at them for the next two million years—will not stop them. Though today's scientists list them as an extinct species, *Homo erectus* never really went away. Their existence never ended in a complete sense. They are in you and every other human alive today. We continue their long run.

It is important to mention the Australopithecines, an extinct genus that included several species. They were not human but certainly were critical to the human story. These upright and bipedal primates evolved in Africa around four million years ago and thrived for some two million years before becoming or giving way to *Homo habils* and *Homo erectus*. Their brains were small, about one-third the size of ours. Males weighed aroundt 90 pounds (42 kg) and were 4-foot, 11-inches tall on average. Females were approximately 64 pounds and 3-feet, five-inches tall. Donald Johanson discovered the fossils of an *Australopithecus aferensis* in 1974 in Hadar, Ethiopia. Nicknamed "Lucy", the find ranks as one of the greatest in science. *Aferensis* may have been the direct ancestral species to *Homo*. Fossils representing hundreds of individual Australopithecines of various species have been found to date in Africa.[270]

BETWEEN WORLDS. Small brained, but upright and regularly bipedal, the australopithecines were fascinating primates who lived for some four million years and were a critical phase in the evolution of humans. The two figures shown here are part of a display at the American Museum of Natural History in New York City. *Photo by the author*

STORIES OF A FORGOTTEN PEOPLE

The accomplishments of *Homo erectus* are remarkable, the stuff of genius and epic stories. They survived for two to perhaps two-and-a-half million years. This alone makes them an extraordinary people. Anatomically modern humans like you and I have been around 200,000 or 300,000 years.[271] This is only ten to twelve percent of the time they lived. And it's doubtful that we will ever exceed their long tenure. More likely, at some point over the next two million years, modern humans will either be gone—made extinct by an all-out nuclear war, supervolcano, or maybe an asteroid strike. A more hopeful scenario would see us evolving into something significantly different, perhaps a human-machine hybrid. In the future a "human" might even be all machine with digital consciousness. It's also possible that we will remain completely organic but evolve into a bizarre space-faring species with an assortment of traits necessary for life beyond Earth. Two million years is a long time for our genus. Relative to other lifeforms, we tend to move and change fast.

According to the best evidence available now, *Homo erectus* were prehistory's great inventors and trailblazers. These were the first hominins known to make and use stone tools extensively; first to control fire; to cook food; to build significant shelters; and, perhaps most impressive of all, the first to expand beyond Africa and live throughout Eurasia in many different and challenging environments. *Homo erectus* were our critical ancestors, the game changers who put us on a new and special path. They made us unique primates by conquering the dark and dangerous night. They took humankind from frightened prey to apex predator status. Unfortunately, entire cultures don't fossilize, only some tangible bits and pieces of them do, and those tend to be very difficult to find. As a result, there are many remaining uncertainties about *Homo erectus*. We don't know details about any possible art, religion, language, or political hierarchy some of them may have had.

How did we get here? is a question of such importance to our self-awareness that the best evidence-based answer to it must be contemplated by all thoughtful, modern people. And no answer to this question can be near competent or complete if it doesn't include the story of *Homo erectus*. Yet most people know little or nothing about them.

Every person alive today owes *Homo erectus* a debt of gratitude for not only surviving and pushing those all-important genes forward, but also for ushering in and establishing forever the survival-by-culture lifestyle that would become the basis for human existence. Their brains were only about two-thirds the size of modern human's brains but these were no fools. As hunter-gatherers-scavengers, *Homo erectus*, more than any other people, transformed humankind from reactive to proactive, from hunted to hunter, from instinctive to intellectual.[272] They were the first humans to cease from cowering in the dark and hiding up in the branches of tall trees. *Homo erectus* were the first people to own the night.

Homo erectus were problem solvers who overcame immense challenges. According to the best evidence, they were the first hominins to control fire, cook food, and build substantial shelters. Their primary tool, the Acheulian stone hand-axe, was in use for more than a million years and helped drive the human revolution. These pear-shaped multipurpose tools had a six- or seven-inch cutting edge and were used to work with wood, plants, and to butcher animals that had been hunted or scavenged. Some archaeologists speculate that they may have held symbolic significance, too, perhaps as the earliest art.[273] Acheulean hand-axes seem to have been crafted with more attention to symmetry and aesthetics than would have been necessary for practical purposes. They might have even been expressions of individuality or even used as tangible symbols for sexual selection, as in: the better one is at toolmaking, the more attractive one is as a potential mate.[274]

AN EDGE AGAINST NATURE. In use for nearly two million years, the stone Acheulean hand-axe may be the most important and longest used human tool of all time. *Homo erectus* and other prehistoric peoples relied on it to help them survive in diverse environments throughout Africa and Eurasia. Some experts think that the Acheulean hand-axe was so important to early humans that it may have, at some point, become the first "money" or tradable commodity.[515] *Photo by the author*

Homo erectus pulled off what is arguably the greatest adventure of all time. Over many thousands of years, on foot and perhaps by raft as well, small independent groups of *Homo erectus* migrated beyond the African continent to settle vast stretches of Europe and southern Asia. They reached as far down to what is today Indonesia. Prolific tool makers and explorers, they endured for an incredibly long time. Civilization accounts for less than one percent of the time *Homo erectus* spent walking the Earth. No one in the twenty-first century can claim to know their species without knowing these remarkable and successful prehistoric people.

ONE ORIGIN OR MANY?

One of the most fascinating and important questions paleoanthropologists grapple with is *how many* origins modern humans had and *where* it all happened. The Out of Africa versus Multiregional Origin debate raged for decades. In question was whether modern humans evolved in Africa and then spread throughout the world (Out of Africa) or evolved separately in Africa and in Eurasia from *Homo erectus* populations that had previously dispersed from Africa (Multiregional Origin). Today, the Out of Africa model looks like the winner, due to the strong fossil and genetic evidence that support it. It is not settled, but most experts today agree that anatomically modern humans most likely emerged in Africa around 200,000 to 300,000 years ago and then, from about 100,000 to 60,000 years ago, began walking, wading, and rafting their way around the globe in a series of migrations spanning thousands of years.[275] Science never sleeps, however, and there are many gaps in the evidence trail so this remains a fluid story. An important question being asked today, for example, is whether we had just one origin in Africa or multiple origins on that continent. In just the last two decades, fossil finds in southern and northern Africa have complicated the old timeline. Anatomically modern humans may have evolved independently in different parts of Africa before merging together as one species. It would be nice if scientific discovery always cooperated with our desire for quick and easy answers. But, of course, reality is under no obligation to be simple and convenient for us. All we can do is work hard to reveal it and then be brave and honest enough to accept what we find.

FROM FORAGER TO FARMER

For most of human existence people were nomadic hunters and gatherers guided by a practical need to eat and drink water. They stayed put if there was food and moved on if there wasn't. And then, as mentioned briefly in chapter two, humankind began the momentous transition to a new survival strategy. Someone, perhaps ten or fifteen thousand years ago, or maybe earlier, did something remarkable. She or he pushed a seed into the dirt with purpose. That person had a conscious plan or intent for a specific plant to sprout from that seed. With patience and a little human help along the way, maybe in the form of watering and weeding, it would germinate and grow, to be harvested and eaten later. This is an idealized event, of course. It is not likely that there was a single inventor of agriculture. It probably happened in multiple places at different times. Remember, these were highly intelligent humans, with brains as large or larger than ours today.[276] There is no doubt that they had sufficient brainpower, plus plenty of awareness of the natural world around them, to figure out the connection between seeds and plants.

Around the same time, some people, maybe the same people in some cases, built new relationships with one or more animal species. The process of domestication, not just taming but fundamentally changing animals to serve human needs, became increasingly popular over the last ten thousand years or so as well. To be clear, the transition from hunting and gathering to farming and pastoralism was uneven and occurred over thousands of years. There is "firm evidence" that as earlier as eleven or twelve thousand years ago people in the Fertile Crescent of the Middle East were modifying local plants.[277] This was thousands of years before there is clear archeological evidence of domesticated grains. In 2017 an international team of scientists from Harvard Medical School, the Hungarian Academy of Sciences, and the Max Planck Institute for the Science of Human History published results of their work in *Nature* showing genetic evidence of Eurasian hunter-gatherer populations co-existing and slowly assimilating with farming populations in the same regions. There was "continuous gene flow" between them over many centuries.[278] Regardless of the details, these "events" were of profound importance for humankind.

The first groups of people who ceased their wanderings, settled down, and invested everything into a few plant and animal species did more than anyone else ever has to build the future we now inhabit. No invention, no great war, no system of government, not even writing, nothing else changed us the way the rise of agriculture did. An important question to consider from our perspective in the human story is whether or not we should we be thankful that it happened. Did the first farmers of prehistory do all the many billions of people who followed them a big favor? There was a time when the answer to that question would have been clear and obvious to everyone. After all, who in their right mind would choose the constant struggles of tiny units of people scavenging dangerous new lands and depending on nature for everything? What is that compared with the glories and comforts of civilization? Thanks to the work of archaeologists, however, we now understand that the answer is more complex.

A BITTER HARVEST?

Somewhere along the way during my childhood I was taught that the agricultural revolution was more than an important cultural transition for humankind. It was nothing less than our salvation, our critical step out of the jungle toward the gleaming civilization on a hill. According to my middle-school teachers, planting crops, raising goats, pigs, and cows elevated us from the miserable existence of chronically hungry and desperate scavengers to well-fed and long-living farmers. Life was so good and food so plentiful down on the old prehistoric farm that human population growth exploded. Moreover, people now had so many calories and free time that they could finally get busy writing, inventing, and building the modern world that we now have the good fortune to live in. But is this true?

One of the more remarkable revelations from archaeology is that human health *declined* when people transitioned from hunting and gathering to agriculture. But this doesn't seem to make sense. If agriculture ignited a population boom, one that continues to this day, then how could it have been bad? Doesn't more people equate to success? Clearly farming increased human survivability. Yes, settling down in one place to depend on planted crops and domesticated animals was positive in that it allowed many more people to eat and therefore live. But the existence of more people isn't everything. What about health, happiness, leisure time, security, and life span? An important aspect of evolution that many people are mistaken about has to do with purpose. Many believe that evolution is about going somewhere with intent, "getting better", and making progress. It's not. Evolution is about survival or extinction, in the present and in the current environment. Evolution is not a ladder of advancement that lifeforms ascend to achieve improved states. A beneficial trait today might be a death sentence tomorrow. Many people think of our large and powerful brains as some kind of pinnacle of evolution, the crowning accomplishment of life on Earth. But what if humankind uses its big brains to invent weapons that bring about human extinction? Should that happen big primate brains would obviously not be a progressive feature of evolutionary success.

Evolution is an unintelligent, indifferent force of nature that acts upon genes within a population. There is no concern or accommodation for comfort and happiness if those things don't impact gene transmission. Pushing DNA forward one more generation is evolution's only measure of success or failure. That's it. The rise of agriculture and animal domestication may have been an evolutionary success for *Homo sapiens* because it allowed many more people to live long enough to make many more babies. Regarding the *quality* of life, however, it has been a mixed bag.

Trading the typical highly diverse hunter-gatherer diet for a very limited one came with a price. Biotechnology expert and plant researcher Denis J. Murphy offers this observation: "The more restricted diet of many farming cultures led to a series of vitamin deficiencies that severely impacted on their well-being and lifespan."[279] A typical early farmer ate a lot more starchy carbohydrates and less animal protein than a typical hunter-gatherer. The cost of having a more consistent food source meant a significant drop in nutrition quality. By one measure, this resulted in a 22 percent erosion of overall health.[280] The farming life made people shorter, too. The average male hunter-gatherer height during the Paleolithic (pre-agriculture) era was just under five feet, ten inches. Late Neolithic men, farmers, were a little over five feet, three inches on average.[281] Settled agriculturalists saw a dip in life expectancy compared with their hunter-gatherer predecessors, as well. A study of eastern Mediterranean Paleolithic hunter-gatherers found that males lived on average 35.4 years and females 30.0. (The gender difference was due to childbirth complications.) By the late Neolithic, when virtually everyone there was dependent on agriculture, life expectancies fell to 33.1 for males and 29.2 for females. Even as late as the nineteenth century, life expectancies had inched up only to 40.0 and 38.4 in that same region.[282]

THE HUMAN WAY. For almost all human existence people have lived in small tribal communities. Civilization—with its farming, urban environments, and thousands or millions of people sharing relatively close living spaces is very new. Civilization accounts for less than one percent of the approximate 2.5 million years of human existence. *Photo of Maasai man in East Africa by the author*

Today's average lifespans are significantly longer than prehistoric hunter gatherers because of very recent advances in evidence-based medicine rather than the mass consumption of wheat, rice, and livestock. To be fair, medical science might never have developed to the extent it has without the agricultural revolution taking place. But our present longer life spans certainly were a long time coming and the kind of comfort and security that many people in wealthy societies are accustomed to today are still not universal. Many millions of people suffer and live short lives now, lives that would seem in many ways to be less desirable than that of a successful hunter-gatherer twenty or thirty thousand years ago.

New and deadly diseases have been a constant problem stemming from the agricultural revolution. Those diseases capable of infecting humans and some wild or domesticated animals are called zoonoses. During the agricultural revolution more and more humans began living in close, intimate settings with domesticated animals. As a result, the rates of these zoonotic diseases skyrocketed.[283] It was an ideal environment for these diseases to thrive because humans and other mammals such as goats, pigs, and cows are relatively very closely related and therefore provide a similar host environment. It also helps that most microbes evolve at extremely fast rates and therefore can evolve to exploit new opportunities in relatively quick order. The toll in human suffering and death over the last several thousand years is so immense as to be incomprehensible—and the carnage continues today. According to a report by the International Livestock Research Institute, an estimated *2.2 million people die each year* due to just thirteen zoonoses.[284] That's a rate of 22 million deaths per decade, 220 million deaths per century, and 2.2 billion per millennia. Those same thirteen diseases cause an additional *2.4 billion illnesses* every year. Approximately 60 percent of all current human diseases are zoonotic and around 75 percent of all new or emerging diseases are as well.[285] One study found that 20 percent of all human illness and death in the least developed nations was caused by zoonoses.[286]

Ecologist/geographer/biologist/anthropologist Jared Diamond is the author of several notable books, including *Guns, Germs, and Steel*; *Collapse*; and *The World Until Yesterday*. He has called humankind's agricultural revolution "the worst mistake in human history".[287] Diamond rejects farming as a glorious leap forward because, he believes, it has failed and continues to fail at providing enough people a better way of life. He points to the staggering levels of economic struggle, malnutrition, disease, and suffering that are still common today—much of it hidden from or ignored by some of the more fortunate members of contemporary humanity. Diamond wrote the following in a provocative essay for *Discover* magazine:

> Forced to choose between limiting population or trying to increase food production, we chose the latter and ended up with starvation, warfare, and tyranny. . . . Hunter-gatherers practiced the most successful and longest-lasting life style in human history. In contrast, we're still struggling with the mess into which agriculture has tumbled us, and it's unclear whether we can solve it. Suppose that an archaeologist from outer space had visited Earth and tried to explain human history to his fellow spacelings. He might choose to illustrate the results of his research by means of a 24-hour clock on which one hour represents 100,000 years of real past time. If the history of the human race began at midnight, then we would now be almost at the end of our first day. We lived as hunter-gatherers for nearly the whole of that day, from midnight through dawn, noon, and sunset. Finally, at 11:54 p. m. we adopted agriculture. As our second midnight approaches, will the plight of famine-stricken peasants gradually spread to engulf us all? Or will we somehow achieve those seductive blessings that we imagine behind agriculture's glittering facade, and that have so far eluded us? [288]

The popular image of dim prehistoric hunter-gatherers groping about the wilderness in misery until benighted agriculturalists finally planted seeds and cultivated utopia has been exposed. The lives of pre-agriculture people may have been tough, but life on the farm wasn't much easier and, in some ways, it was much harder and less healthy. Yale University political scientist and anthropologist James C. Scott cites an abundance of archaeological evidence opposed to this view in his book, *Against the Grain: A Deep History of the Earliest States*. He writes: "Contrary to earlier assumptions, hunter-gatherers—even today in the marginal refugia they inhabit— are nothing like the famished, one-day-away-from starvation desperados of folklore. Hunter-gatherers, in fact, have never looked so good—in terms of their diet, their health, and their leisure. Agriculturalists, on the other hand, have never looked so bad—in terms of *their* diet, *their* health, and *their* leisure."[289] Scott sees our big transition as more than planting crops and domesticating animals. It was about planting people in one place and domesticating *them*, too.

The act of settling down and farming led to land ownership, rulers and workers, wealth and poverty, slavery and war, all on a grand scale. Of course we cannot condemn in full the agricultural revolution and the civilization it enabled because so much good has come from them. If we want to understand how humankind got to this point, we can't make the mistake of viewing the agricultural revolution as a wonderful event with only positive results. But neither can we idealize or romanticize prehistoric hunter-gatherers.

As a product of and current member of modern civilization, I am likely influenced by bias, of course, but I find it difficult to imagine small and scattered populations of prehistoric hunter-gatherers ever matching the best accomplishments of the last couple of thousand years or so of agriculture-based civilization. Remember that, with some exceptions, hunter-gatherers stayed on the move, maintained very small populations, and did not accumulate much in the way of possessions. Would some or all of them found their way to a path that would later produce microscopes and telescopes, brain surgery and spaceflight? Without agricultural tying them down to a permanent settlement and providing the food surpluses necessary to swell their numbers, I struggle to see it.

THE HUNTER-GATHERER LIFESTYLE

Many retired professional athletes say that more than anything they miss the close bonds they shared with teammates. For many, the camaraderie of the locker-room is as important and memorable touchdowns scored or championships won if not more so. Military veterans commonly describe feelings of trust and loyalty to their fellow soldiers as their primary motivation to fight and risk their lives when combat began. There is something special about being in a small group with agreed-upon goals that feels right and important to us. And so it should, because this is who we are. It was only a moment ago, that we all lived in social units of perhaps a hundred others. We knew everyone, and everyone knew us.

People today seem most psychologically and physically comfortable, healthy, and productive when they belong to groups no larger than about 150 members, which happens to be the average size of a hunter-gatherer group.[290] "Modern humans, in effect, evolved to be social machines that produce and refine ideas," writes geneticist and anthropologist Spencer Wells in his book *Pandora's Seed: The Unforeseen Costs of Civilization*, "and perhaps this explains why management studies show that people seem to work best in the kind of small, focused teams where this ancient hunter-gatherer process can take place."[291]

The typical daily environment for many people today includes many strangers, near-constant noise, toxins, and streams of distractions. All of this may be taking a terrible toll on the minds of modern people. Wells thinks there is a problem: "This psychological mismatch between the densely populated, noisy agricultural world and the sparsely populated hunter-gatherer is almost certainly one of the reasons for the psychological unease felt by many people." Mental illness currently is one of the leading global health burdens. Measured in some ways it is *the leading* cause of disability worldwide.[292] Brandon Hidaka, a medical doctor who researches nutrition science, asserts that more modernization correlates with higher lifetime risks of mood disorders. "Mental and physical well-being are intimately related," he writes in his study of depression as a disease of modernity. "The growing burden of chronic diseases, which arise from an evolutionary mismatch between past human environments and modern-day living, may be central to rising rates of depression. Declining social capital and greater inequality and loneliness are candidate mediators of a depressiogenic social milieu. . . . More money does not lead to more happiness. By appealing to evolutionary proclivities, like a desire for energy-dense food and status competition, the economic and marketing forces of modern society have engineered an environment promoting decisions that maximize consumption at the long-term cost of well-being. In effect, humans have dragged a body with a long hominid history into an overfed, malnourished, sedentary, sunlight-deficient, sleep-deprived, competitive, inequitable, and socially-isolating environment with dire consequences."[293]

What about the perceived value of a human life in a prehistoric hunter-gatherer society versus modern society? It might have been better, given the strong human need for social interactions and relationships. The stress of life without grocery stores and refrigerators would be difficult for most modern people, of course, but think of the constant feelings of high value and self-worth that came with membership to a small prehistoric hunter-gatherer group. The individual was a crucial factor in *everyone's* survival. These were people who constantly relied on one another, communicated with each other, and shared everything. In such living conditions, it is difficult to imagine prehistoric people having self-esteem problems.

Compare this lifestyle with that of most people today. I don't know about you, but if I were to die right now, neither the world, my nation, nor my city would skip a beat. My death would be virtually unnoticed. It would be business as usual. Sure, my family members and a few friends might shed a tear or two, but their survival would not suddenly be at greater risk because I was gone. They would still be able to find food at a grocery store, lock the doors at night, and move on with their lives. In prehistory, the death of a typical person would have had much more impact. All humans lived, *for more than two million years*, in small, intimate social units that placed a high value on everyone's life. Perhaps, then, we might expect that many people today would feel out of place and psychologically stressed, in part at least, because they coexist with large numbers of strangers who care little if anything about them. In a sense, the typical human who was just recently in prehistory something like an essential hero is now something close to invisible and irrelevant.

The positive pushback against all this for the individual may be as simple as regularly spending time in quiet, natural environments such as a park, forest, or beach. Finding and joining a social unit of some kind might be helpful, too. I suggest a small, goal-oriented group with difficult shared challenges of some kind. Maybe combine the two needs and join or form a hiking, birdwatching, or stargazing group. It might be good if power is somewhat evenly distributed within the group, too. And strive to make sharing and communicating a standard feature of your new tribe. The goal is to recreate your own hunter-gatherer band as an answer those genetic rumblings within your human cells. It might only take one night per week at the bowling alley or a club bike ride on weekends might help make life in today's world a little more agreeable.

AGRICULTURE AND INEQUALITY

Archaeological evidence shows that inequality of wealth and power was very low in typical hunter-gatherer societies but rose dramatically once farming and pastoralism gained momentum. Social/political/economic inequality is significant because it impacts many aspects of people's lives in a society. It has also led to terrible spasms of violence throughout the human past.[294] Tim Kohler, a professor of archaeology and evolutionary anthropology at Washington State University was lead researcher of a 2017 study on inequality that looked at more than sixty prehistoric sites. "People need to be aware that inequality can have deleterious effects on health outcomes, on mobility, on degree of trust, on social solidarity—all these things," he said in a statement released by Washington State University. "We're not helping ourselves by being so unequal."[295] Societies that are more equal tend to have higher levels of trust and people are more willing to assist each other, Kohler added.

Inequalities set in motion by the agricultural revolution some ten thousand years ago, are still present today. In the United States, for example, the three richest people in 2016—Bill Gates, Jeff Bezos, and Warren Buffett—owned more wealth than the entire bottom half of the American population combined. That's more than a 160 million people or 63 million households. America's top 25 billionaires together now hold more than $1 trillion in wealth. Globally, eight men own the same wealth as half the human population, 3.6 billion people, according to a 2017 Oxfam report.[296] Such statistics represent more than some kind of real-life Monopoly game. Higher degrees of inequality are connected to life expectancy rates. For example, the richest 1 percent of American men live 14.6 years longer on average than the poorest 1 percent of men. The difference for women is 10.1 years on average.[297]

Stanford University history professor Walter Scheidel studies the formation of states and economic inequalities. One of his books, *The Great Leveler Violence and the History of Inequality from the Stone Age to the Twenty-First Century*, details some of the problems that were created by the transition from hunter-gatherer societies to agricultural societies and how they continue to cripple us in many ways today. Scheidel told me that he doesn't view this great shift to be a mistake necessarily because it is not as simple as

WHAT HATH WE WROGHT? Civilization has been wonderful and devastating to human health and wellbeing. The transition from hunting and gathering to farming generated food surpluses, a population boom, and specialized labor. This resulted in elevated levels of creativity and economic productivity. Agriculture-based societies have enabled people to create magnificent art and walk on the Moon. But at what cost? Civilization also has been the source of vast human suffering, global ecological destruction, and it could ultimately lead to the extinction of *Homo sapiens*. *Photo of Petra, Jordan by the author*

one survival lifestyle being *completely* superior to the other. "Settling down to farm created all kinds of problems," he said, "from
despotic government and endless war to vast economic inequality,
and every time new means were invented to address these problems
they also created new dangers: prosperity-driven climate change
and environmental degradation and the possibility of nuclear war
are merely the most recent examples. Had our distant ancestors
continued to hunt and gather, we wouldn't have to deal with any of
this. But in that case we'd all still be hunter-gatherers, and what a
waste of our brain power that would be."[298]

Scheidel has deep concerns about humanity solving the more
serious problems farming dependence brought us. "Looking back at
thousands of years of history, I am not terribly optimistic about this.
In the past only violent shocks had real impact. It is true that the
past does not determine the future and this century (not to mention
later ones) might just turn out to be totally different. But current
trends point towards greater rather than reduced inequality –
open-ended automation, the aging of populations coupled with
mass migration, potential fall-out from climate change, and the
final frontier of genetic and cybernetic enhancement of the human
body. That's not an environment in which greater fairness is going
to arrive peacefully."[299]

The best we can say about agriculture is that it has been a mixed
bag for humankind. Yes, it caused, or allowed, the rise of complex
civilization. Stratified societies filled with specialized workers set
up a wonderful and never-ending explosion of writing, mathematics,
science, and art. Yes, agriculture's bounty of food, though nutrition-
ally costly to our health in several ways, enabled a population boom
that continues to this day. And, yes, it is likely that humans would
not have achieved as much in science, music, literature, and art,
or invented vaccines and computers if not for farming. Agriculture
was a critical factor in how we got here and presently it is the only
way billions of people can be fed. But some of its results have been
deeply troubling and remain key challenges for us to overcome.

THE PRIMATE WHO SPOKE

Once we began putting our big brains to use for something more than moment-to-moment survival we became a force the world had never seen before. We imagined and built upon existing ideas and knowledge to enhance our power and abilities. Modern humans are notable for being highly intelligent and habitually social. Our closest relatives, the other great apes, are also impressively intelligent. But our massive, neuron-dense brains are three times larger than any other living primate. Despite the popularity of strange beliefs and the violence we so often commit against nature and against ourselves, we reign as the supreme thinkers of this planet. We are capable of symbolic thinking and future planning to such a level that it makes us unique among all Earth life. And communication in the form of spoken language is the key; it is the primary reason we have been able to make so much of our thoughts and dreams. Complex oral and later written languages give us the ability to think outside and beyond one or a few brains. With words, we can harness the intelligence of millions of minds across time.

Language is the fuel and framework of complex human culture. It enables us to share information and retain it so that societies can accumulate and amplify knowledge. There is no way for me to visit and chat with Henry Thoreau, Gandhi, or Martin Luther King, Jr. because they long gone. But they can still communicate to me. Thanks to complex oral and written language, I am able to read thousands of words they left behind on the pages of books. I can know their ideas, hopes and fears—even their personalities to some degree. Today we can build on thoughts that emerged from human minds literally thousands of years ago. We can be inspired by ancient people and their stories. Language connects us not only to people in the present but throughout history as well.

The complex language we rely on today likely developed long ago in prehistory from simple calls to groupmates. Unfortunately, very little about the origin(s) of language can be declared with much if any certainty because spoken words don't fossilize, uttered sentences don't freeze, and nobody wrote down the tale because writing came many thousands of years after oral language. But it's easy to understand the functionality of auditory communication of one kind or another. We see it all over the animal kingdom. Nature is loud.

It is easy to imagine our primate ancestors warning peers about a predator nearby. This behavior would have obvious selection value, with those groups who were better at sharing and receiving such vital information in real time having higher survival and reproduction rates. The more detail packed into an oral warning about a nearby lion's precise location the better the chances that a hominin or two in the group avoid being eaten. Ververt monkeys in Africa today have at least ten specific vocalizations for predators that allow them to distinguish between, for example, an approaching eagle or snake.[300] You wouldn't want to duck when you should have jumped. But when did the first prehistoric people, maybe even more distant ancestors, use something complex enough that we can consider it language? That's a big question and there is still no satisfying scientific answer.

Many favor a view shared by the late evolutionary biologist Stephen Jay Gould and noted Massachusetts Institute of Technology professor Noam Chomsky, "the founder of modern linguistics". The claim is that language is an innate feature of modern human brains and only arose relatively rapidly and recently, at some point between 70,000 and a hundred thousand years ago.[301] In short, language is just too special to be the endpoint of routine and slow evolutionary processes. It must have been the spectacular side-effect of an unlikely convergence of genes, brain expansion, and changing brain structures. There has been a lot of fascinating work done with *FOXP2*, the so-called "language gene".[302] In conjunction with some other genes, it is believed to be a prerequisite to complex language. So maybe language is the product of a fortunate genetic quirk, one that had an immediate impact and allowed us to soar as the planet's greatest ever communicators. Humans today have the *FOXP2* gene and Neanderthals did too. Unfortunately, no *Homo erectus* DNA has ever been salvaged that might reveal if it carried the gene. Perhaps language emerged as an unintended consequence of human brain evolution. When the human brain evolved to facilitate better memory and more complex socialization, for example, perhaps the stage was set within for speech to occur.[303]

LET'S TALK ABOUT TOOLS

An intriguing idea about the origin of language is that tool-making may have been a critical factor. A telling archaeological experiment using modern people showed that novices become better faster at making stone tools via oral instruction more so than by imitation or emulation.[304] When early hominins sat around striking stones to craft their tools, it would have benefited them to be able to communicate in detail about the process. There would have been selection pressure to use increasingly complex oral communication for easier and more efficient transference of tool-making techniques from experts to novices and from one generation to the next. Tool making might have sparked or at least provided a big boost to the human language revolution.

As far as the timing of it, I side with the experts who advance the idea of language showing up much earlier in the human story than a hundred thousand years ago. I think it more likely that it came into use two million years ago—at least. No one knows, but it would not surprise me if some Australopithecines had been speaking a surprisingly impressive language or proto-language as far back as three million years ago. Sadly, it is unlikely we can ever know if this was the case. I favor the idea that language is extremely old and was directly subject to the forces of natural selection. It seems clear that a group with superior communication abilities could have done more and survived at a higher rate than other hominin groups. Consider the remarkable accomplishments of *Homo erectus*, the intelligent, social, and creative beings who walked and rafted their way, over many generations, from Africa all the way deep into Europe and Asia, and possibly further. I cannot imagine them doing all of this without some kind of sophisticated communication ability. Maybe they relied heavily on hand gestures, prehistoric sign language to share information.[305]

Homo erectus went where no one had gone before. Arising at least 1.9 million years ago in Africa, these people figured out how to succeed and survive for some one-and-a-half million years.[306] They explored and solved immense challenges across diverse and dangerous continents. Try to imagine the natural environments they encountered during more than a million years of migrations. *Homo erectus* eyes were the first human eyes to see the towering trees, vast deserts, beaches, lakes, wetlands, rainforests, mountain ranges, and snow-covered lands of prehistoric Eurasia. They imagined, planned, and experimented their way across a good portion of the planet. *Erectus* hands crafted tools, controlled fire, and built shelters. Could they have done all that—given the necessary planning and ingenuity—while relying on nothing more than a few calls and grunts to communicate with each other? Did *Homo erectus*, arguably the all-time greatest travelers and explorers, first great hunters, and possibly the first seafarers sit around their campfires at night and just stare at one another in silence? I doubt it.

Daniel Everett, author of the book, *How Language Began: The Story of Humanity's Greatest Invention*, describes them as nothing less than "lords of the planet"[307]. He writes: "*Erectus* did not simply walk single-file or run randomly around the world. They were organized. They were smart. They were a society of cultured humans. And they must have had language."[308]

In a paper they co-authored, language researchers Stephen Pinker and Paul Bloom wrote that it is generally unappreciated "how important linguistically-supported social interactions are to a hunter-gatherer way of life. Humans everywhere depend on cooperative efforts for survival. Language in particular would seem to be deeply woven into such interactions, in a manner that is not qualitatively different from that of our own 'advanced' culture."[309] And *Homo erectus*—the most enduring human species of all time and the first to breach the borders of Africa—were hunters and gatherers. Constant and heavy dependence on communication would have been standard for them. Pinker and Bloom point out that language likely would have been present in *Homo habilis* groups, too, before *Homo erectus* existed. "[H]uman language, like other specialized biological systems, evolved by natural selection," they conclude. "Our conclusion is based on two facts that we would think would be entirely uncontroversial: language shows signs of complex design for the communication of propositional structures, and the only explanation for the origin of organs with complex design is the process of natural selection."[310]

WHAT HAVE WE DONE?

A Human Timeline

- **The universe begins** 13.8 billion years ago, providing the time, space and matter necessary for humans to exist.
- Some 4.6 billion years ago, **the Sun forms**, providing the energy human life will later depend on.
- **The Earth**, future home of *Homo sapiens*, forms 4.54 billion years ago.
- The **ultimate origin** of humans is unknown but likely traces back to the beginning of life on Earth which, based on fossil evidence, is estimated to have occurred approximately four billion years ago.
- The **last common ancestor** shared by modern chimps and modern humans lives in Africa perhaps as recently as five or six million years ago or perhaps as far back as thirteen million years ago.[311]
- The earliest members of the genus *Homo*, **the first humans**, belong to a species named *Homo habilis* ("Handy Man") living in eastern and southern Africa approximately 2.5 million years ago.
- It is unknown when hominins first began using an **oral language**. Some scientists think that it happened within the last 80,000 years or so. Others think it happened well before the evolution of anatomically modern humans, with some African *Homo erectus* populations being among the first people to speak.[312]
- *Homo erectus* learns to **control and use fire** in Africa.
- *Home erectus* is the **first human species to migrate beyond Africa**, making it all the way to Europe and Asia.
- The first anatomically modern humans **evolve in Africa** around 200,000 to 300,000 years ago.
- For more than ninety-nine percent of human existence people will live as **semi-nomadic hunters-and-gatherers**.
- Possible evidence of early **symbolic thought** shown in the form of geometric patterns carved into an African rock dated to about 100,000 years ago.[313]
- Anatomically modern humans **migrate beyond Africa**

between 100,000 to 60,00 years ago.

- *Homo sapiens* mate with, kill, and/or outcompete **Neanderthals** in Europe.
- People **create hand stencils and paint animals** some 40,000 years ago in Sulawesi, Indonesia. **Geometric shapes** are carved into rock at Wharton Hill, Australia perhaps 36,000 years ago.
- People paint the **Chauvet cave art** more than 30,000 years ago in what is today southern France. More examples of art begin showing up around the world.
- The rise of **agriculture** is underway twelve thousand years ago.
- People **come together, organize, plan, design, and build Gobekli Tepe**, a stone complex in Turkey, more than 10,000 years ago.
- Humankind is well into its **transition to civilization** by six thousand years ago. As more people are living in permanent locations (cities) and relying on farming and domesticated animals to live, societies are becoming more complex, with byproducts of social and wealth inequality, labor specialization, rulers and elites, state warfare, and organized religions.
- **Writing** begins more than five thousand years ago.
- Sumerians use, possibly invent, a **number system** approximately five thousand years ago. Ancient Babylonian and Egyptian cultures develop mathematics.
- About 4,500 years ago, **people in South America build Caral**, a city in what would later become Peru.
- **Egyptian, Greek, and Roman empires** flourish and will influence human cultural development for millennia to come.
- **Aristotle, Archimedes, Plato, Pythagoras, Hippocrates, Galen, Socrates**, and others help lay groundwork for modern science.
- People in Mesoamerica **develop writing and show astronomical sophistication** 2,500 years ago.
- **Alexander the Great** conquers a vast empire around 330 BCE.
- Buddhism, Christianity, Confucianism, Hinduism, Islam, Judaism, Shintoism, Sikhism, Taoism, Zoroastrianism develop toward becoming **popular organized religions**. All told, humankind will create tens of thousands of religions that will claim the existence of millions of gods.
- **Mongol conquests** (1206-1368) kill more than thirty million people.

- Johann Gutenberg introduces **moveable type printing press** to the West, helping to set the stage for mass publishing and reading in coming centuries.
- Beginning in 1492, the voyages of Christopher Columbus connect Old and New Worlds. The subsequent **European conquest and colonization of Americas** results in more than seventy million deaths, mostly from imported diseases.
- Leonardo da Vinci **sketches a human fetus in the womb.**
- Galileo observes **the moons of Jupiter** through a telescope he built in 1610.
- An estimated twenty-five million killed in **Qing Dynasty war against Ming Dynasty** (China, 1618-1683).
- Isaac Newton's work on **physics and mathematics** in the 1600s helps propel human understanding of reality forward.
- Antonie van Leeuwenhoek develops **microscope technology** in the seventeenth century, studies cells, discoverers numerous microscopic lifeforms.
- Wolfgang Amadeus Mozart composes the musical piece *Eine Kleine Nachtmusik* in 1787.
- The rise of machines and factories marks the beginning of the **Industrial Revolution** (1760-1840).
- **Sanitation practices** improve, preventing countless illnesses and deaths. Scientists advance **Germ Theory** in the 1800s.
- Millions of people captured, transported in chains, and forced into lifelong servitude as **African slave trade thrives** and peaks in the eighteenth century.
- British New World colonies break with England to become the **United States of America** in 1776.
- Charles Darwin's presentation of the **theory of evolution,** *On the Origin of Species* is published in 1859.
- New Zealand grants **women the right to vote** in 1893, the first self-governing country to do so.
- In 1905 Albert Einstein publishes the Theory of Relatively, later to be known as the **Special Theory of Relativity. It reveals the surprising interconnectedness of matter and energy, time and space.**
- More than fifteen million people killed in **World War I** (1914–1918).
- Alexander Fleming discovers penicillin in 1928. Scientists develop **antibiotics** which will save hundreds of millions of lives.

- An estimated sixty million people, most of them civilians, killed in **World War II** (1939-1945).
- **Nuclear fission bombs** dropped on two Japanese cities in 1945.
- James Watson and Francis Crick **discover DNA structure** in 1953.
- Approximately 20 million people die from **purges, neglect, and official government policies** in Soviet Union during the twentieth century.
- The United States detonates the first-ever full-scale fusion bomb in 1952. Thermonuclear warheads, **the most powerful weapons ever,** produce destructive energy via the same process that the Sun produces life-giving energy for Earth.
- Mao Zedong's **Great Leap Forward** (1958-1962) leads to famine and more than 30 million deaths in China.
- **Evidence of the Big Bang**, which happened 13.8 billion years ago, detected in 1964 in the form of cosmic microwave background (CMB) radiation.
- United States and Soviet Union amass **tens of thousands of thermonuclear weapons** in the late twentieth century. This is enough destructive power to annihilate global civilization and possibly cause human extinction.
- Humans complete **nine voyages to the Moon** between 1968 and 1972. The Apollo missions include six landings that place a total of twelve people on the surface of the Moon.
- The **complexity, power, and popularity of computers soars** in the late twentieth century.
- **Human genome** sequenced in 2003.
- The **Internet and World Wide Web** used by millions before close of the twenty-first century. In 2017 more than half of the world's population is online and some three billion people are active on **social media**.
- Governments and corporations invest billions of dollars into **Artificial Intelligence** research and development in the early twenty-first century.
- In the second decade of the twenty-first century, more than 15,000 scientists in 184 countries sign the **"Warning to Humanity"** document that highlights numerous negative environmental trends that threaten human well-being. These include: climate change, deforestation, loss of fresh water sources, species extinctions dying oceans, and human population growth.

By at least 30,000 years ago art was becoming a standard feature of human culture. *Photo by the author*

Ancient Egyptian civilization, beginning approximately 3100 BCE, was one of the first large and complex agricultural-based societies. It lasted for more than 3,000 years. *Photo by the author*

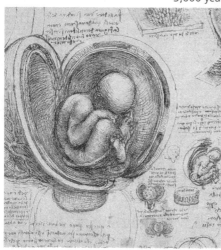

Leonardo da Vinci sketched the human fetus around 1510.

A Saturn V rocket lifts off on its way to the Moon. *Photo NASA*

FINDING THE PRESENT IN THE PAST

Scientific thinking and the technology it produces, coupled with the remarkable creativity and courage humans are capable of, enable us to achieve extraordinary feats. I have had the privilege of conducting long interviews with special people who have been to the extremes of humanity's physical reach. I spoke with one of the three people who visited the bottom of the Marianas Trench, the deepest point of the ocean. I also have interviewed five astronauts who voyaged to the Moon and back, as well as one who lived in space for 132 days aboard the Mir space station. Thinking about those men now, as representatives of human ability, inspires pride in my membership to a surprising and storied domain, kingdom, class, order, family, genus, and species. We are the once few and fragile apes who climbed down and stood up so long ago. With feet to carry us far; hands to craft tools from wood and stone; and big brains to remember, plan, talk, and imagine anything, we found ways to not only survive but thrive. Our story of survival and success is beautiful and deserves a prominent place in the mind of everyone everywhere. Neither our embarrassing hunger for inefficient and dangerous delusions, the insane self-destruction of so many pointless wars, nor short-sighted behavior toward our planet's life-sustaining ecosystems has overwhelmed our greatness, not so far. In a relatively short time we have gone from taming fire to solving many mysteries of both the atom and universe. Of course we stumble well short of anything close to perfection and a consistent show of wisdom. But no one may yet reasonably declare us utterly stupid or all bad. Humankind is far from a lost cause, especially so early in the story and with so much potential still swirling within those potent and in some ways limitless brains.

Imagine the world without us. Several human species came and went at various times over just the last couple of million years. Any one of them might have endured instead of us. Or, *all* hominins could have failed. Change an environmental variable here, fail to come up with a clever idea there, maybe a random genetic mutation happens, or one never happens, and Neanderthals would have made it through to the present instead of us. The slightest tweak to the environment could have meant no hominins of any kind survive and it's the arthropods and bacteria who reign. Imagine if more than a billion years ago Eukaryotes just never evolve. There is no critical fusion of Archaea and bacteria; no complex multi-celled organisms; and no plants or animals, ever. It could have gone like that or a trillion other ways. But we're here; we made it. You belong to a winning team, so far.

As science can best tell it at this moment in time, the human story can leave us feeling unsatisfied, confused, and frustrated. There are gaps in the information, contradictions, and uncertainties. But what we can know right now matters. Make the best of what has been revealed and pay attention as more comes to light. Scientific discovery is under no obligation to cooperate with our desires or for our convenience. And that's okay. Paleoanthropologists, archaeologists, historians, and others will continue working this challenge. They keep questioning who we are and work hard to find the evidence that builds real answers. It may help to remember that often the most exciting stories of all are unfinished and unpredictable, those likely to have surprising twists and turns ahead. Having at least some of the story is superior to blindness and confusion. A partial and imperfect scientific answer beats a complete but made-up answer every time. Consider the definition and description of humankind to be a work in progress. The story of *Homo sapiens* is like a spectacular and (mostly) beautiful painting, a potential masterpiece, though one with a lot of white space on the canvas remaining to be filled in. Every new scientific discovery is a new stroke of a paintbrush, one more clue toward discovering ourselves and answering how we got here.

Humankind is not the history of one nation, the claims of one religion, or the traditions of one culture. It is much older and larger than that. The story of *Homo sapiens* began as far back as the first organic entities on this planet at least 3.95 billion years ago.[314] There is an unbroken chain of life from then to now. It links every living human to those initial molecules that started it all. Acknowledging the wonder of this and knowing some of the details as revealed by science is essential for self-awareness. Science's best answers to *What is life?* and *When did life begin?* are indispensable to the explanation of who we are.

Every person grows from knowing as much as possible about the path to now. Understanding the depth of time and gaining perspective on the events and processes that shaped us can change one's view of self and everything else. A change, I feel, that is for the better. Learning that there have been many different human species—and all but one have gone extinct—means acknowledging that the success and survival of *Homo sapiens* was not and is not assured. It alters how we think of ourselves when we feel the immense depth of time our species spent moving across Earth's surface as hunters, gatherers, and scavengers, as opposed to staying in one place tending crops and animals. No one need feel obligated to return to a completely prehistoric lifestyle, of course, but why not mix in some helpful elements of the human past with the best of the human present? Awareness of how we spent more than 99 percent of our existence is valuable and practical knowledge. Knowing who we are now requires knowing who we were in the past and how we got here. The human story continues to be written with many chapters ahead of us. As we move forward, I hope we are smart enough to include and make us of more scientific awareness of self.

Chapter 7

WHY DOES THE RACE CONCEPT FAIL?

Sometimes history, culture, and the human mind conspire to create and maintain a seductive lie packaged as an "undeniable truth." One of humanity's most tragic and enduring self-inflicted wounds is the concept of biological races. This is one of those "undeniable truths" that is neither obvious nor true and collapses in the light of science and reason. Race beliefs are fueled and fortified by cultural momentum, subconscious biases, gaps in scientific and historical knowledge, and perhaps a bit of intellectual apathy. The belief leads billions of people to "know" with great confidence that distinct biological races exist, and that nature or supernatural forces have sorted all humans into. However, as we will see in this chapter, the popular versions of race belief are illogical, invalid, and unsupported by evidence. Becoming aware of this, to finally perceive and conceive of humankind as it is, can not only enhance the life of an individual, but also might better our chances of achieving more cohesive and cooperative societies. The title of this book is *At Least Know This*. Given our past and present problems with division and discord, understanding something about the reality of human biological diversity certainly is essential, minimal knowledge.

This chapter is not about unscientific claims of one race having mental or moral superiority over another race or any of the other standard tenets of racism. Such problems are inevitable byproducts of the race concept itself, which is the foundational problem addressed here. The biological races that most people believe in have come not from nature but from culture. There is real biological diversity in humankind, of course, but the fabricated categories of popular race belief neither describe nor organize it accurately. Billions may believe in biological race groups, but they are made up entities that do not exist anywhere other than in our human imagination.

Although the names and number of races vary across societies and over time, the mistake is the same. Generation after generation, children are taught to believe in whatever "commonsense" and "plain-to-see" race claims and rules their specific culture currently maintains. After that, confirmation bias goes to work behind the curtain of consciousness to reinforce the belief over a lifetime. The result of all this are big troubles for humankind in the form of more social and psychological divisions in a global population with more than enough of those. The simple explanation is that *Homo sapiens* is too young and genetically similar a species—too biologically tangled—to naturally fragment into the discrete, consistent, and meaningful race categories people believe in.

It is difficult for many to see the flaws in race belief because they recognize that *Homo sapiens* is not a species of clones. Though all people on Earth today are about 99.9 percent genetically identical[315], there is significant and visible biological diversity. Some individuals and some populations, on average, are taller or shorter than others; some have darker or lighter skin than others; some have one kind of hair type while others have a different type. And, though unreliable, such visible traits can sometimes be linked to relatively recent prehistoric ancestry in certain geographical regions. This is because, during prehistory, some people in some environments evolved traits that were different from those of some populations in other environments. But all this means in the end is that humans have inhabited different environments and some degree of biological diversity exists. It does not justify the claim that meaningful, distinct, and consistent biological race categories can be discerned. Anatomically modern human populations were not apart from one another in different environments long enough to evolve into subspecies. Again, all people everywhere today remain *99.9 percent identical*. This can be hard to accept given our penchant for obsessing on and exaggerating the significance of relatively superficial and trivial differences such as nose-width, hair texture, and skin color, but this is reality. It also is common for people to confuse cultural behaviors—such as language, dress, and religious belief—with genetic destiny. Regardless of whatever race someone might be assigned to, anyone can be raised from infancy within any culture to successfully learn and use the language, as well as understand and take part in all customs, the economic and religious system, and so on.[316]

Race belief matters because it distorts the reality of who we are. Everyone can benefit and attain a sharper worldview by recognizing its inherent irrationality. A devout race believer can never know humankind because she or he constantly views it through an imaginary prism. The confusion the race concept sows impacts how not only overt racists, but also how all people think of and interact with others. Remember, racial identity is supposed to be inherited, a genetic destiny no one can ever fully escape. Race defines and cages vast groups of people in a way that inspires false knowledge about their presumed potential abilities and limitations, as well as their morality, trustworthiness, and threat risk to others. What makes this such a senseless and frustrating problem is that it is based on bad logic and a gross misreading of the human story. So much so, that it could not survive without childhood cultural indoctrination, steady social reinforcement, and a pandemic of ignorance regarding prehistory, history, and current biology. Fortunately, nothing more than a bit of science and critical thinking easily exposes race as a bogus belief built out of wrong assumptions and maintained by false walls.

THE EMPEROR HAS NO RACE

It can be difficult to cease "seeing" something that is not there when everyone else around us claims to see it. For some, irrational race belief is part of a worldview they have grown comfortable with. There is a lot tied up in it, and emotional investments don't go down without a fight. Public popularity and backing from large and respected institutions such as governments, school systems, and religious organizations also help race belief thrive. This is because the human subconscious mind is a sucker for majority consensus as well as power. As you think about the race concept, remember that all people, you included, are highly vulnerable to falling in with the herd and obeying authority figures with little thought, even when doing so doesn't make sense. These are standard human features, nothing to be ashamed of, but definitely nothing to ignore, either.

Some readers may be cringing in discomfort at this point or, more likely, scoffing at the suggestion that popular race belief is an unscientific delusion. To these readers I only ask that you continue reading. Please try your best to keep an open and honest mind open as you encounter the facts and ideas ahead. I promise that there is no stealth political agenda hiding in this chapter. I am not pushing a well-intentioned lie in hopes of alleviating the horrors of racism. I am not so blinded by a desire for people to get along that I can't see reality.

A SIMPLE QUESTION

One clue that something is very wrong with popular race belief is that no one can even agree on how many races there are. Numbering and naming biological race groups should be easy, given the claimed existence of objective and natural borders that fragment human-kind. But there is not today nor has there ever been consensus on this. It doesn't matter who you ask, scientists or laypersons, you are sure to get conflicting answers on this simplest point. The problem is that *Homo sapiens* just does not accommodate the structure of popular race belief. Augustin Feuntes, chair of the Department of Anthropology at the University of Notre Dame and a National Geographic Explorer, explains that the tight genetic kinship of all modern humans renders popular race claims not only inconsistent and arbitrary but biologically trivial:

> Neither genetics, nor behavior, nor height, nor body, face, or head shape, no skin color, nor nose, what type of hair, nor any other biolog-ical measure divides modern humans into subspecies. If you compare the genetic differences between any two humans from anywhere on the planet, they are much, much smaller than those between any two chimpanzees from Eastern and Western Africa. It is a stunning fact. Humans are spread across the whole world and chimpanzees are found in only a relatively tiny swath across the center of Africa, but humans are far more genetically similar to one another. This pattern is the same for most all comparisons between humans and any other mammal—we are among the most genetically cohesive and most wide-spread of any animal on the planet, a combination that is amazingly rare in the animal kingdom.[317]

Carl Linnaeus, the influential classifier of life, named and described the characteristics of four races, or "varieties" of modern humans back in the eighteenth century. It is not difficult to recognize some subjective bias behind his thought process: Africans ("crafty, negligent, governed by caprice"); Americans (Native Americans, "Obstinate, merry, and free, regulated by customs"), Asians ("severe, haughty, governed by opinion"); and Europeans ("gentle, inventive, governed by laws").[318] Though influential, Linnaeus's racial groupings and accompanying horoscopes of humanity didn't stick as he wrote them. They were modified and rewritten ad nauseum by people who proposed different races, fewer races, and more races. The only thing that has been consistent over the last 400 years or so is that no scientific breakdown of humankind into biological race groups is universally accepted by all experts. It is difficult to imagine a consensus remaining so elusive if racial categories were real, observable, measurable, and objective as claimed by believers. Like many ideas that spring wholly from human minds, the race concept is infinitely flexible because it owes nothing to evidence and has no basis in reality.

The same absence of consensus that has always existed among and between biologists, anthropologists, sociologists, historians, and geneticists extends to the public as well. During the time I spent researching and writing my book, *Race and Reality: What Everyone Should Know About Our Biological Diversity*,[319] I asked just about anyone who would talk to me about their race beliefs. While I was not surprised that no common number of races emerged among laypersons, I was fascinated by how people went to one of two opposite extremes to justify their answer. Some would struggle with the maddening challenge of trying to sensibly shoehorn more than seven billion living humans into three or four tightly defined race categories. Others would go in the opposite direction and keep adding categories to accommodate Australian Aborigines, residents of the Indian subcontinent, Arabs, Pacific Islanders, and so on. Neither tactic ever worked, of course, as believers inevitably ended up with little confidence in their declared number, even if they typically still clung to their belief in the race concept.

How many oceans are there? This is another question I often raise when discussing races. Those who did well in geography class are quick to answer "five". But is that correct? Sure, there are five named oceans: the Pacific, Atlantic, Indian, Arctic, and Southern. But this is a misleading answer. "Five" is the *cultural* answer. It's the number based on names and divisions that people made up. The answer from *nature* is that there is but "one". Anyone who doubts this can inspect a globe and immediately see that there is just one big ocean with no borders. It's most clear when viewed from the bottom. Typically, a child is taught early in life to believe in multiple oceans and then spends the rest of his or her life never doubting or questioning the "obvious reality" that there are five distinct oceans, when there is only one in the natural, objective sense. Anything seem familiar here? The reason the number of races has varied so widely, from three to sixty or more[320], is that no objective, consistent, and sensible natural markers exist by which one can segment humankind in a way that coincides with popular race belief.

THE PEOPLE WHO STUDY PEOPLE DON'T BELIEVE IN RACES

A study published in 2017 sought to find out what anthropologists think about the concept of race. The data revealed a "dramatic rejection" among professionals in all subfields. "We observed consensus that there are no human biological races," the researchers wrote.[321] They added that there was, however, a common recognition that races do exists as cultural or "lived social experiences" and as such can have important health effects on individuals.

The following are highlights from the American Anthropology Association's official "Statement on Race" published in 1998 (italics added for emphasis):

- [B]oth scholars and the general public have been conditioned to viewing human races as natural and separate divisions within the human species based on visible physical differences. With the vast expansion of scientific knowledge in [the twentieth century], however, it has become clear that *human populations are not unambiguous, clearly demarcated, biologically distinct groups.*
- Throughout history whenever different groups have come into contact, they have interbred. The *continued sharing of genetic materials has maintained all of humankind as a single species.*
- As they were constructing US society, leaders among Euro-

pean-Americans fabricated the cultural/behavioral characteristics associated with each "race," linking superior traits
with Europeans and negative and inferior ones to blacks
and Indians. *Numerous arbitrary and fictitious beliefs* about
the different peoples were institutionalized and *deeply
embedded* in American thought.

- *Racial myths bear no relationship to the reality of human
capabilities or behavior.* Scientists today find that reliance
on such folk beliefs about human differences in research has
led to countless errors.
- [We] now understand that *human cultural behavior is
learned*, conditioned into infants beginning at birth, and
always subject to modification. No human is born with a
built-in culture or language. Our temperaments, disposi-
tions, and personalities, regardless of genetic propensities,
are developed within sets of meanings and values that we
call "culture."
- The "racial" worldview was *invented* to assign some groups
to perpetual low status, while others were permitted access
to privilege, power, and wealth.
- Given what we know about the capacity of normal humans
to achieve and function within any culture, we conclude that
*present-day inequalities between so-called "racial" groups
are not consequences of their biological inheritance but
products of historical and contemporary social, economic,
educational, and political circumstances.*[322]

EVERYONE CAN'T BE RIGHT

One telling clue that the concept of race is an intellectual train wreck
is that different societies have contradictory race beliefs. Around
the world, cultures have conflicting and often irreconcilable race
rules and claims. While it all may seem to make perfect sense to
people within a given society, it's clear when viewed from the global
perspective that some race believers, or maybe all race believers,
are wrong. This alone should go a long way toward eroding the overwhelming confidence in race beliefs that is so common today. But it
doesn't, unfortunately, because most people know very little of other
cultures and their beliefs. Most people grow up in one society, absorb

the customs, values, traditions, beliefs and rules, both written and unwritten, and never bother to peek over the fence. It can be jarring, then, when one learns that their race belief system—sold to them as obvious, objective, natural, and universal—is just one of many cultural creations.

Brazil, for example, has very different criteria for who is black than the United States does. In the Cayman Islands, where I lived for several years, race rules are different from both Brazil and the United States. I had many Caymanian friends who were not identified as black people at home but were often identified as black people when they flew to the United States for a visit. Such a strange and dramatic change can be easily explained if racial identification is just a cultural game. But this is a deep mystery if races are determined by biology and ancestry, as most people believe. Does anyone really believe that flying from home at an altitude of 35,000 feet and landing in a different country somehow alters an individual's biology and rewrites his or her historic and prehistoric ancestry?

America's one-drop rule is based on a crude idea of contamination that says a child with parents of different races must be given the racial designation of the parent with the lower racial status.[323] This, for example, would mean that in most cases throughout American history the offspring of a black parent and a white parent would be assigned to the black race because the black race was viewed as inferior to the white race. The one-drop rule has existed in American culture as both law and social norm since at least 1662.[324] While this custom might seem acceptable, logical, and natural to many Americans, even today, it's hardly universal to race belief as most likely presume. Haiti, for example, has its own one-drop rule. I was friends with a light-skinned Haitian woman in college. She was consistently viewed as black in the United States but somehow turned white every year when she returned home for Christmas and summer breaks. This happened because in Haitian culture, even minimally visually detectable hints of white ancestry in skin color, facial features, and hair, can make a person white. This is America's one-drop model reversed. So, who gets it right, Haiti or the United States? They can't both be correct. One must be right and the other wrong. Or, there is that third option again, of course. Both could be wrong in the sense that they are promoting cultural whimsy disguised as natural reality.

WHEN WERE RACES REAL?

Time also makes a mockery of race belief. In the minds of many people, races are what they are because they're based on observable and measurable traits. The race concept is comparable to the Earth revolving around the Sun. That's just the way it is; the Earth and Sun share the same gravitational relationship today that they did a hundred or a thousand years ago. And human races are no different, we are assured. The race concept is not a trend or fashion that is susceptible to human quirks, according to the claim. The reality, however, is that, just as there is no consistency of races today across cultures, there has never been consistency across time. Anyone can look back at history and see clearly that race is a hodgepodge, crazy grab-bag of arbitrary claims and made-up rules. A thought experiment: imagine gathering up a few random "white Irish" and "white Eastern European" people today. Lure them into your time machine and send them back to New York City of 1810. When your time travelers arrive, they will soon discover that their race has changed or at least degraded en route. Irish people and Eastern Europeans, identified simply as white today, were somehow not-quite white back then. They certainly didn't suffer the same violence and indignities that black people did in America at the time, but they were not granted the status of "regular white people" with all due benefits. Different time, different rules.

There is neither a definition nor a formula for race that has proved itself to be timeless and universal. Race belief is constantly changing, never logically constructed, nor consistent across cultures. It is a fabricated phenomenon that answers not to nature and genes, but to the whims of every culture it burdens.

CAN'T WE JUST SEE RACE?

Perhaps the most common of all defenses of irrational race belief is that one can "just see it". Race is in the faces of everyone around us every day. Who can't recognize the difference between a dark-skinned Kenyan standing next to a blonde Finn or Swede, right? The reason this justification works so well in the minds of race believers is that they are working from a false perception of humanity. Humankind is not one dark-skinned Kenyan and one light-skinned Swede standing side-by-side. Real biological diversity cannot be reliably measured, separated, and sorted by skin color and hair type. The reason I can't present the idea of a seven-foot tall person standing next to four-foot person and convince anyone that there are two distinct human races, the tall race and the short race, is because people are well aware that literally billions of unseen height variations fall between those two extremes. For this same reason, an imaginary police lineup example of two or three selected "race representatives" standing side-by-side should fail, too. Skin color, facial features, and hair type exist over a wide range of gradients across humanity. They fall upon a long spectrum that offers no natural borders to support the popular racial categories. A black African standing next to white Swede, who is next to a Japanese person, only seems to validate biological race categories because literally billions of people have been omitted from the scene. It is a contrived and dishonest mental image that renders the majority of the human population invisible and unaccounted for.

Contrary to common belief, we cannot see race in every face. At best one can make a reasoned guess about someone's recent ancestry, and some are better than others at it, but this is not nearly as easy to do as people imagine. Give me the world population to work with and I could fool race believers all day long. I saw many people during my travels in Syria, Jordan, and Egypt, for example, that were almost perfect visual matches for people I knew in the United States and the Caribbean. Without clothing, language, and behavioral clues, race becomes more difficult if not impossible to spot in the faces of a strangers.

Most people lack a realistic view of human biological diversity because they have encountered so little of it. Unfortunately this doesn't stop them from *imagining* that they know the human species very well because of who they know within their society and who they have seen on TV, in movies, and online. There is no substitute, however, for close encounters with the actual human variation that exists in the real world. Outside of tourist hotels and tour buses, the walls of race crumble fast because one finds contradictions at every turn. Dark skin, for example, is a primary determinant of membership in the black race. But I saw many very dark-skinned people in India and Australia who would not be labeled "black" by most race believers. The darkest people I have ever seen anywhere lived in Papua New Guinea. I have also met Fijians who could easily pass as African-Americans or black Caribbean people, based on their hair, facial features, and dark skin color. And yet indigenous Fijians are about as genetically distant from African-Americans as anyone on Earth can be. (Keeping in mind that *all* humans are very closely related genetically.) I've seen some people in villages up in the Andes Mountains who, if they were dressed in neutral clothing and did not speak, could easily fool many typical race believers into thinking they were Japanese or Chinese people. I have met a few Native Americans also who with little effort could fool almost any observer into thinking they were Thai, Chinese, or Japanese.

People who think they can easily identify their culture's particular biological race categories with a mere glance at the face of anyone they encounter can only think this because they either haven't seen enough of the world's people or, if they have, an irrational belief is warping their perceptions. The high degree of geographical and social isolation that is common to most people can lead anyone, regardless of intelligence or education, to imagine all of humanity as a handful of caricatures. But real humanity defies such a simple classification scheme. It is important to be aware that human minds are forever trying to assist their owners by finding shortcuts. Serving the goals of speed and efficiency, the mind will cling to an image—perhaps a single face, body shape, or even a behavior—and then use it as *the* factor for lumping millions of people together in a "biological" group. Thanks to confirmation bias, one can then cling to it forever and feel confident that it makes sense, regardless of how many contradictions are encountered over a lifetime.

Although the race concept is often described in scientific language, the truth is that science offers it no substantial aid and comfort today. It may be possible to analyze certain genes or select combinations of physical features to identify more closely related populations or clusters of people. But this inevitably becomes an exercise that is too subjective and inconsistent to support traditional race beliefs. Researchers *Luigi Luca Cavalli-Sforza, Paolo Menozzi and Alberto Piazza addressed this in their book The History and Geography of Human Genes:* "At no level can clusters be identified with races, since every level of clustering would determine a different partition and there is no biological reason to prefer a particular one. . . . From a scientific point of view, the concept of race has failed to obtain any consensus; none is likely, given the gradual variation in existence. It may be objected that the racial stereotypes have a consistency that allows even the layman to classify individuals. However, the major stereotypes, all based on skin color, hair color and form, and facial traits, reflect superficial differences that are not confirmed by deeper analysis . . .[325]

Awareness is the key here. Instead of expecting and looking for differences, divisions, and justification for races in the face of every passing stranger, we might recognize our real similarities and connections to all modern humanity. Those Native Americans I mentioned having met were not Asian, but they certainly displayed faint echoes of their ancestors' prehistoric migrations to the New World from Asia. Every person alive today is the product of epic journeys of survival and every one of them leads back to a common human beginning. All journeys therefore, can be thought of as a single shared voyage. Seeing family everywhere you go on Earth and being reminded of a shared past at every turn may seem to some like the worst kind of over-romanticizing, but it is reality. It's science.

PHYSICAL/BIOLOGICAL ANTHROPOLOGISTS ON RACE

Physical and biological anthropologists work in a subfield of anthropology that specializes in human biological diversity, both prehistoric and current. Given their focused study of bones, blood, and genes, it is reasonable to suspect that they might know as much or more about race than anyone. The following are excerpts from the official "American Association of Physical Anthropologists Statement on Biological Aspects of Race" (italics added for emphasis):

- All humans living today belong to *a single species, Homo sapiens,* and share a common descent.
- Much of the biological variation among populations involves *modest degrees of variation in the frequency* of shared traits. Human populations have at times been isolated, but have *never genetically diverged enough* to produce any biological barriers to mating between members of different populations.

- There is great genetic diversity within all human populations. *Pure races*, in the sense of genetically homogenous populations, *do not exist* in the human species today, nor is there any evidence that they have ever existed in the past.
- There are obvious physical differences between populations living in different geographic areas of the world. Some of these differences are strongly inherited and others, such as body size and shape, are *strongly influenced by nutrition, way of life, and other aspects of the environment.*
- The *only living species* in the human family, *Homo sapiens*, has become a highly diversified global array of populations. The geographic pattern of genetic variation within this array is complex, and presents *no major discontinuity.*
- Humanity *cannot be classified into discrete geographic categories with absolute boundaries.* Furthermore, the complexities of human history make it difficult to determine the position of certain groups in classifications. Multiplying subcategories cannot correct the inadequacies of these classifications. [Modern humans are too blended genetically to be sensibly divided into biological race groups. Creating more racial groups ad infinitum will not eliminate this reality.]
- Partly as a result of gene flow, the hereditary characteristics of human populations are in *a state of perpetual flux.* [Gene flow is the movement or sharing of genes between two populations]
- There is no necessary concordance between biological characteristics and culturally defined groups. On every continent, there are diverse populations that differ in language, economy, and culture. There is *no national, religious, linguistic or cultural group or economic class* that constitutes a race. [Race belief is often supported by a common misperception that culture (learned behaviors) is biology (inherited traits).]
- Racist political doctrines find *no foundation in scientific knowledge* concerning modern or past human populations.[326]

GENETICS VS. RACE

In 2010 the results of the first DNA sequencing of an indigenous hunter-gatherer from the Kalahari Desert were published in the prestigious journal *Nature*.[327] The researchers analyzed the genes of five southern Africans and the data delivered another blow to race belief. It turned out that two of the southern African tribesmen had a high degree of genetic difference between them. So much so that— even though they looked similar and had lived their entire lives within walking distance of one another—they were *less related* to each other than either of them was to a typical white European or Japanese person. Let that sink in for a moment. These are two men who any typical race believer would not hesitate to place in the black race because of their common skin color, hair type, geographical proximity, and African ancestry. Yet, according to their genetics, the men logically should be in a race group with white Europeans or Asians rather than share one together. Remember this study the next time you hear someone claim that races are obvious and easy to see.

Some people claim that modern genetics has vindicated or proved that traditional race categories (Whose tradition?) are real, objective, and biological. But the rapidly advancing field of gene study has not done this. In fact, many geneticists are outspoken about the need to abandon race as a means of analyzing and describing *Homo sapiens*. Charles N. Rotimi, chief and senior investigator for the National Human Genome Research Institute and director of the Center for Research on Genomics and Global Health, believes that race clouds thinking and stands in the way of fully appreciating and maximizing gains from research on the human genome. Rotimi makes this point:

> We all have a common birthplace somewhere in Africa, and this common origin is the reason why we share most of our genetic information. Our common history also explains why contemporary African populations have more genetic variation than younger human populations that migrated out of Africa [approximately]100,000 years ago to populate other parts of the world . . . Our evolutionary history is a continuous process of combining the new with the old, and the end result is a mosaic that is modified with each birth and death. This is why the process of using genetics to define "race" is like slicing soup: "You can cut wherever you want, but the soup stays mixed".[328]

Former *New York Times* science writer Nicholas Wade has been trying to make a scientific case for race for many years. In his 2014 book *A Troublesome Inheritance: Genes, Race, and Human History* he writes, "Even when it is not immediately obvious what race a person belongs to from bodily appearance, as may often be the case with people of mixed-race ancestry, race can nonetheless be distinguished at the genomic level."[329]

Identifying a gene or two that may be linked to a prehistoric or historic population in a specific geographical area, however, does not validate biological races as they are widely believed in today. Everyone has both recent and distant ancestors somewhere. Who gets to decide which genes, geographical locations and time period defines the races? Look back far enough and everyone has the same ancestors. Follow every trail back far enough and it ends in Africa. Matching a genetic marker within the genome of some person in Detroit, Michigan with West Africans, for example, does not establish the existence of a meaningful and consistent entity called the black race because that person in Detroit and all West African people are likely to be far less related to millions of others identified as black people.

Wade's book, although popular with enthusiastic race believers, drew negative reviews from several experts and, most notably, a letter of rejection and condemnation signed by 139 of the world's leading population geneticists.[330] The common race categories don't work because the human species has never cooperated with anyone's desire for neat and consistent *biologically based* categories that can contain hundreds of millions of people with some respectable degree of coherence and consistency. Geneticist Alan Templeton has studied the concept of race for decades and concludes that it does not rise to the level of valid science. "Humans show only modest levels of differentiation among populations when compared to other large bodied mammals, and this level of differentiation is well below the usual threshold used to identify subspecies (races) in non-human species," Templeton states. "Hence, human races do not exist under the traditional concept of a subspecies as being a geographically circumscribed population showing sharp genetic differentiation. . . . Genetic surveys and the analyses of DNA haplotype trees show that human 'races' are not distinct lineages, and that this is not due to recent admixture; human "races" are not and never were "pure." Instead, human evolution has been and is characterized by many locally differentiated populations coexisting at any given time, but with sufficient genetic contact to make all of humanity a single lineage sharing a common evolutionary fate."[331]

Science shows us that humankind has been in a "a state of perpetual flux" throughout the past and remains so today. Drawing the arbitrary borders of traditional biological races between and around populations where none exist and calling it science is the height of irrationality and blind arrogance. Neither our distant nor recent past supports this. The genes inside of our cells right now do not validate anything close to popular race belief.[332] Race is a fantasy that obscures our vision. It is one of the key obstacles preventing us from knowing ourselves.

RACE AND DISEASE

The idea of "race diseases" seems to come up in every discussion that challenges the reality of races. This common claim says that there are race-specific diseases and they help prove the existence of biological races. This is wrong, however, because diseases do not conform to the particulars of a society's current belief in biological races. And it leads to serious problems. Cystic fibrosis, for example, is underdiagnosed in dark-skinned people because many doctors think of it as a "white" disease.[333] Sickle cell disease is often cited and described as a "black disease" in the United States. But it's not. Sickle cell disease doesn't believe in biological races either, nor does it follow anyone's race rules.

Sickle cell disease is a problem of abnormal hemoglobin in red blood cells. Normal red blood cells are donut shaped but sickle cells are rod-like and inflexible. This causes difficulties because they can get stuck in vessels, which slows or blocks the flow of blood. This disrupts the transport of oxygen in the body and can cause pain or even death. The disease is not race-specific. It is an evolutionary adaptation to Plasmodium, the malaria parasite. Sickle cell is a problem that impacts many people around the world who have ancestry that is connected to regions with high rates of malaria in the past. Race is not the issue. For example, the trait is relatively common in central India, Nepal, and the Middle East. A small town in Greece, Orchomenos, has one of the highest rates of sickle cell anemia in the entire world, twice that of the African-American population.[334] And the disease is virtually nonexistent among black South Africans.[335] Even in the United States, where it is commonly thought of as a "black disease", eleven of every twelve black people have no biological connection to it.[336]

THE STRANGE CASE OF BARACK OBAMA

Former US president Barack Obama is one of the most famous African-Americans of all time and popular perceptions of him present a good example of not only the absurdity of race belief but also how it is dangerous in healthcare. Obama's father was a Kenyan and his mother an American of European descent. Obama is brown-skinned and has a black parent, so he was designated a member of the black or African-American race early in life, as is the custom in America. So while he may be culturally belong to the black or African-American race, what about his biology? With his immediate ancestry being white European and *East* African, Obama is relatively far removed from an overwhelming majority of US blacks or African-Americans. Most African-Americans have recent ancestry from *West* Africa. This means Obama is likely to be very distantly related to most African-Americans. (Africans have the greatest genetic diversity because people have lived on the continent the longest.) Ironically, his white mother's genes likely would give Obama his closest biological link to other African-Americans because of the intermating that has occurred in US history with gene flow travelling mostly from the white population to the black (one-drop rule).

Let's pause here and recall that the biological race categories so many people believe in are supposed to be about biology. Race, we are told, is in the blood; it's based on the objective measures of an individual's genetics and kinships. Race is supposed to be biological ancestry. But Obama is in a different biological race than that of his own birth mother. Such is the cognitive fog that race belief depends on to endure.

All this may be interesting enough, but imagine the serious consequences if a doctor were to allow Obama's race to influence diagnosis and treatment. Some risk factors might be ignored while others are given too much attention. Obama could be prescribed a race-specific drug, controversial already, but worse perhaps when so far off the mark of a patient's actual biology.[337] Visible traits such as skin color and facial features are inadequate clues for making medical decisions, and yet doctors rely on them all the time.[338]

Race is a problem in medicine and healthcare because it is a terrible substitute for biological knowledge and a risky shortcut to diagnoses. No doctor or medical researcher can glance at a patient and know for sure everything relevant about the last several thousand years of the person's ancestry. So why do they do it? Probably because doctors are human, too, and many of them have an irrational belief in the biological race categories of whatever society they happen to have been raised in. The reality is that a doctor can't even be sure about the parents and grandparents of a patient without some investigation. Race belief is too haphazard and arbitrary to trust in influencing medical decisions. And yet it shows up every day in many doctors' offices and hospitals around the world.[339] In my book *Race and Reality: What Everyone Should Know About Our Biological Diversity*, I give the example of an eight-year-old boy who was brought to a hospital and almost put through unnecessary exploratory surgery. Because he looked "white" or "European" to the doctors, they did not recognize that his symptoms pointed to acute sickle-cell disease which is what he had. The doctors believed, as many people do, that sickle-cell is a racial disease that effects black people so it was simply not on their mental diagnostic radar. Fortunately, the doctors caught their mistake just before cutting the child open.[340]

Tay-Sachs is another disease often cited by race believers as some kind of proof that races exist somewhere other than in human minds. This genetic disorder is rare in the general population with 1 in 200 people carrying the recessive gene. However, 1 in 27 Ashkenazi Jews are carriers.[341] A baby is at risk of inheriting the deadly disorder when both parents are carriers. But there are significant problems with the claim that Tay-Sachs shows biological races make sense. For one, Judaism is a religion that people can join. Jews do not normally proselytize or recruit newcomers like some other religions, but outsiders do join, and non-Jews have been joining for millennia. A college friend of mine converted from Christianity to Judaism before marrying his Jewish finance. I'm confident that neither his blood, genetics, nor ancestry changed at the completion of his classes with a rabbi or at any point during or after the wedding ceremony. He certainly didn't instantly develop a greater chance of becoming a carrier of the Tay-Sachs gene. Fortunately, I'm happy to add, proactive testing has virtually eliminated Tay-Sachs disease from the Jewish population.[342]

French Canadians in southeastern Quebec and Cajun people in southern Louisiana have had their problems with Tay-Sachs disease. So too have the "Pennsylvania Dutch", a small, semi-isolated US population with a lifestyle similar to that of the Amish.[343] Does this mean these groups are races too? It should be obvious that a genetic disorder like Tay-Sachs is the possible consequence of a small population with historically restrictive mating practices, nothing more. If a few thousand Manchester United football fans in England or a small population of Star Trek enthusiasts in San Diego were to begin mating primarily within their respective groups, a genetic disease could emerge to threaten their newborns at a higher rate than those babies born in the larger, more genetically diverse population outside those artificial mating restrictions. This alone would not, however, sensibly justify believing in the existence of the Man U. and Trekkie races or subspecies.

SPORTS AND RACE: THE GREAT MIRAGE

For all the talk about fostering unity and bringing people together, various sports have long served as powerful promotional agents of divisive race belief. Black sprinters and basketball players; white coaches and team owners; smart white quarterbacks handing the ball to aggressive black running backs, and many more images act as steady drips of bogus evidence that validate many of the unfounded claims about race that bounce around in the minds of fans. Billions of eyes watch sports events every year and see what they believe when it comes to race. It is remarkable, for example, that "white men can't jump" is a common belief when the reality is that white men and white women have dominated Olympic high jump competitions.

The Olympics, NFL, NBA, and so on are vast playgrounds where misperceptions, confirmation bias and other subconscious shenanigans wreak havoc on unprepared minds. The common tendency to mistake cultural influences and present behaviors for ancient genetic destinies runs rampant among casual sports observers, fans, athletes, and coaches. The reality, however, is that sports do not confirm the claims of race belief, but rather offer a clear and invaluable window into the processes by which such an irrational proposition can flourish.

Let's consider competitive running as an example of a sport that shows how people misinterpret evidence against race belief as evidence for it. Running is a universal ability that was selected for by evolution, developed, and honed long ago in prehistory. It is not difficult to imagine, a million years ago, one *Homo erectus* racing another to the nearest tree for fun, status enhancement, or simply to get first pick of the fruit. The ancient Greeks held foot races more than two-thousand years ago and track and field (or athletics) is globally popular today. Running may be the oldest, purest, most intricately measured, recorded, and analyzed of all human physical abilities. It is also a common source of confusion for defenders of race belief who see it as powerful proof for the reality of biological races. "Just look at the Olympics," I've been told countless times. "Black people win all the gold medals in track and field; the 100-meter final is nothing but black guys. Black runners win all the distance events, too."

This is supposed to show that there really is something to this race thing because we can see with our own eyes dramatic differences in performance. There is no affirmative action or other such social manipulation on the track to corrupt the results. The black race produces the fastest runners and we can see the undeniable proof of this in the results. Two things allow this argument to seem compelling to so many people: One is that most people know little or nothing about track and field. The other is that most people know little or nothing about human biological diversity.

First of all, the results on the track are not what they may appear to be. Yes, something is going on as revealed by results, but it's not surprising in the context of a realistic understanding of human diversity and it does not credibly support race belief. Yes, there is subset of humanity that currently dominates the Olympic sprint events. But this subset is not "black people". If blacks are the fastest sprinters due to some kind of racial blessing, then why has there never been a black African gold medalist in the 100 or 200 meters in the entire history of the Olympics? Africa is a giant of a continent with a huge population to draw from. There have been many African Olympic track champions in other events, so we know it's possible for Africans to win races on the biggest stage. And yet, *no black African 100-meter Olympic champion ever*. So why are black Africans so slow? Are Africans somehow less black than the dark-skinned athletes who have been winning most of the Olympic sprints in recent history?

People who watch the Olympics and see "black people" crossing the finish line first in the 100 and 200 meters, fail to recognize that virtually all recent medalists in the Olympic sprints have been members of a very specific and small population of people who share recent West African ancestry. Dark skin and membership to something called the black race are not the relevant factors, as is widely believe. Most male Olympic 100- and 200-meter gold medalist of the last forty years or so has been a black athlete from the United States, Great Britain, Canada, or the Caribbean. No male black African has ever earned a medal of any kind in the 100 meters at the IAAF World Championships.[344] What the top dark-skinned speedsters of today have in common is that they are the descendants of slaves taken from West Africa to Britain or the New World. Their success in the sprint events is most likely explained by some unique combination of cultural/historical factors and possibly some set of genes favorable for sprinting that exist at a higher rate within this dispersed but relatively small population.

We find a similar story with the African runners who have increasingly dominated Olympic distance events after Ethiopian Abebe Bikila's barefoot marathon victory at the 1960 Rome Olympics. Again, race believers see dark-skinned winners and think: *black race equals elite endurance*. But if so, why has there never been an African-American or West African champion in any Olympic race above 800 meters? Are they not black enough? Are they somehow not real members of the biological black race? Why does a typical race believer see a pack of Kenyans leading the Boston

Marathon in the later stages and conclude that "black people" have great endurance? Why not ask instead why black people, with the minute exception of some East Africans, are nowhere to be found in topflight distance competitions? The reality is that, excluding recent immigrants, worldclass black American, Canadian, British, Caribbean and West African distance runners are exceedingly rare to nonexistent. Why hasn't Nigeria, for example, the most populous nation in Africa, ever produced an Olympic medalist in any distance event? Clearly there is something other than dark skin or having African ancestry that is the key to sprint and distance running success at the highest level. But every time race believers see another dark-skinned person cruise to victory in the 5,000 or 10,000 meter races they believe their conclusions about black racial running superiority have been confirmed, not realizing that it has been almost exclusively a small subset of *East* Africans who have excelled in the Olympic distance races. Africa has fifty-four nations and a population of more than one billion people. Most Africans are dark-skinned to some degree and are designated as members of the black race by most race believers. And yet we do not see domination by "black runners" drawn from across the continent. Contrary to the common assumption, biological blackness (whatever that is supposed to be) is irrelevant to success on the track.

Even to say that East African runners are the best distance runners is misleading. The great distance runners to come out of East Africa have come almost exclusively from very small regions within specific nations, mostly Kenya and Ethiopia. In the case of the Kenyans, for example, virtually all have been members of the Nandi and a few other tribes that are collectively known as the Kalenjin people. Despite being a minority within Kenya's national population, these tribes in one area have produced virtually all of Kenya's great runners. Therefore, it is too vague to describe these athletes as simply "Kenyans." Although that is certainly is much less misleading than attributing their success to being simply African or black.

The greatest problem with thinking of all elite dark-skinned sprinters and distance runners as members of a single shared race is that in many cases these runners have *more genetic distance* between themselves than they do to some random Asian or white European person. Despite similar skin color and a connection to the same continent, Kenyans, African-Americans, Ethio-pians, Nigerians, black Brits, and Caribbean black people cannot be placed into a single biological race by any logical or scientific means that makes sense in relation to the rest of the human species. The genetic diversity of these various people is so great as to make their presumed common biological race absurd. For example, Usain Bolt winning the 100 meters and Mo Farah winning the 5,000 cannot be the vindication of biological races that many imagine because Bolt, a Jamaican, has historically recent ancestry to West Africa, while Farah, a British citizen who immigrated to the UK from Somalia as a child, is biologically an East African. Given the relative great genetic distance between West Africans and East Africans the cul-tural process of jamming them into a single biological race together is laughable.

It is important to note the injustice that accompanies the kind of irrational race-based analysis so common throughout sports. The success of black athletes often is believed to be a function of their genetic, inherited, and mostly *unearned* talents. For successful white athletes, however, hard work, determination, courage, and mental ability are more likely to be credited. Again, sports, for those who care to look with open eyes, disproves rather than confirms this harmful stereotype. For example, many fans may believe that Magic Johnson was a great basketball player because he was black and therefore had been born with magic calf muscles or something like that. But the truth is different and readily apparent for anyone who cares to consider it. By all accounts, Johnson was obsessed with basketball from very early in his childhood. He would practice on a court for an hour or more every day *before* the school bus picked him up for elementary school. When he got home from school, he played and practiced for hours more before finally literally sleeping with his basketball.[345] And then, when he began high school, Johnson *really* got serious about the game. Consider that white player Larry Bird, Johnson's primary rival, was deemed by many to have been great because he practiced jump shots virtually every day of his life, as if the six-foot, nine-inch Bird had been born genetically defi-cient, without any innate potential to make it to the NBA much less become a superstar.

Michael Jordan, arguably the greatest basketball player ever, was widely viewed in his prime as a fast, explosive, leaping, high-flying representation of the natural, gifted black athlete who was born to play. But what about his work ethic and constant obsession with winning? Jordan earned the reputation of being one of the most focused and hardest workers to ever play in the NBA. And those great East African distance runners I wrote about previously? In my view, they rank among the toughest and most dedicated athletes in all of sports. I have been to the Great Rift Valley in Kenya where they live and train. What they call routine—the high altitude, brutal hills, sparse diet, and relentless focus—is the stuff of nightmares for other runners. When the perception of an athlete is run through the filter of race belief, the result can be that the athlete is cheated of the respect that he or she deserves.[346] When a sport is racialized and beliefs allowed to alter what is seen, human beings are robbed of their individuality and denied due respect and credit, just as so often happens in the world beyond sports.

WHAT THE TRUTH MIGHT BRING

The premise of this book is that scientific thinking and awareness of fundamental, evidence-based knowledge has the potential to enhance our lives. This is never truer than when we use science to help us recalibrate and improve the way we perceive ourselves and others. Humans have always been deeply social creatures—for better and for worse. We will continue living lives of constant cooperation and competition into the foreseeable future—also for better and for worse. Therefore, how we think of our social connections, current and potential, is profoundly important. Given our problems with irrationality, fear, tribalism, and aggression, it is difficult to imagine social perfection happening any time soon or ever. However, an honest and realistic view of humankind, specifically a scientific perspective that reveals the close kinships and overwhelming similarities shared among ourselves as humans, might make it significantly easier for people to live in peace while accepting the real rather than imagined and fabricated differences between us.

Knowing that biological races are not credible categories of human beings will not solve racism and deliver universal harmony because *cultural* races and many other divisions of our own making do exist and have real-world effects. But seeing ourselves as we really are and robbing the biological race concept of its illegitimate power over us can only help. This chapter is a crucial supplement to the answer presented in an earlier chapter (*Who are We?*). Understanding why the race concept fails is necessary to understanding what it means to be human and a member of *Homo sapiens*. This is a critical awareness that can be good for the world in a practical sense. As the number of people who possess a more logical and evidence-based view of human diversity increases, the better chance we will have of building better communities and perhaps realizing a more sensible, peaceful world.

Chapter 8

WHERE ARE WE?

One memorable night in the Cayman Islands several years ago I was up late, well past midnight, alone in the backyard with my telescope. It was dark, cool, silent, no mosquitos, and my wife and kids were asleep in the house. Even the dog had abandoned me. It was just me and the universe. At some point that night I had recognized the moment for what it was—one of those rare moments of near perfection that sneak up on you every now and then over a lifetime. I remember observing a few fine sights through the telescope that night. But it was with naked eyes that I saw something that stunned me, perhaps even scared me a bit. It definitely changed me forever. For the first time, I saw where I was.

Above me, *close to me*, the entire universe soared by as a spectacle of endless light against infinite blackness. It was not the whole universe, of course, but it may as well have been to me. I never looked through my telescope again that night. There was no need. The stars had come to me. The Milky Way galaxy swirled over me like a frozen, frenzied explosion of pure art. Like a collision of Van Gogh and Pollock paintings, it was a mashup of light and matter on a scale that I struggled to accept as real. For more than an hour I simply stared up. No one was around to laugh at me, so at one point I reached out with a hand and tried to touch it. But, of course, I already had. Every person touches the Milky Way galaxy every moment of his or her life, because every human lives inside of it. Every one of us is a part of it.

I recall at one point feeling slightly disoriented or dizzy during my encounter with the Milky Way. The sky was so clear and dark that it presented my galaxy to me in a way that I had never experienced. All the spinning, orbits, and revolutions suddenly seemed more real than ever. It was my imagination, I'm sure, but I *felt* movement. Standing barefoot in grass and looking up from one little planet, I sensed for the first time that I was on rocky sphere, spinning on its axis at more than a thousand miles per hour, flying around the Sun at some 65,000 mph, and rotating around the center of a galaxy at some half-a-million mph.[347] All this while I was simultaneously soaring through the universe on yet another trajectory. No classroom lecture, book, documentary, or planetarium show had touched me the way that night did. I've never been the same since. I was already interested in astronomy, so it didn't inspire me to care more about space. It didn't change my outlook on life and death or anything dramatic like that. It did, however, enhance my existence by leaving me with a new and deeper appreciation for *where I am*.

Greg Simpson, a 59-year-old Florida native, has owned a dozen or so telescopes over his lifetime, including one he and his wife made from a kit. He became interested in astronomy back in his high school days and hasn't stopped stargazing since. But why? What does he get out of looking up? "It's enhanced my life by exposing me to the entire world of science and how science shows us how the universe actually works, which is simply amazing and fascinating," he said. "What I have gained by looking through my telescopes is a deeper appreciation for the wonder and grandeur of the universe. There's something incredibly appealing in actually being able to see, however imperfectly, some of the objects out there in the galaxy and beyond. I may have observed the Orion Nebula, for example, through the eyepiece a hundred times over the years, but it always seems fresh and new each time."[348]

A THIN ZONE OF LIFE

The first step in attempting to answer the question *"Where are we?"* is to locate ourselves on the Earth. This is not as obvious as one might imagine. Most people, it seems, give this little thought but it matters. Saying simply that we live "on Earth" leaves out too much. Humans live in a relatively small and very specific region on Earth. Knowing this is potentially important because it could change the way many people view our species long-term survivability and our relationship with the rest of this planet's life as well.

Homo sapiens inhabits an extremely small fraction of the overall Earth. The planet's crust is only *one percent* of the planet's total volume but it has been home for all people. Even this is not an adequate answer, however. On land, the crust is about 19 miles (39 kilometers) thick on average. But people do not live throughout the crust, of course. Furthermore, 71 percent of the planet's crust is covered by ocean, rendering it uninhabitable to people without significant technological assistance. Astronauts, submariners, and house-boat residents aside, humankind lives exclusively on top of the land portion of the crust. Therefore, *Homo sapiens* inhabits little more than a quarter of the Earth's thinnest, uppermost layer.

The atmosphere, the layer of gases that surround the Earth, is another key part of the answer. *Homo sapiens* is located *inside the atmosphere*. And for good reason: no atmosphere, no human life. It's about sixty miles thick and gravity keeps the gases down where life needs them. But more than ninety percent of the atmosphere is inhospitable to humans. Without elaborate technological support such as an airplane, spacesuit or spaceship, people are incapable of surviving above the lowest end of the atmosphere due to the cold temperatures and lack of oxygen. There are, however, other micro-bial life forms that are able to survive more than a mile down into the crust and several miles above in the stratosphere.

To humans, down on the surface, the sky may seem immense, perhaps even limitless. Relative to the planet's size, however, the atmosphere is extremely slim, something like the thin peel on a piece of fruit. I have interviewed many astronauts over the years and a couple of them mentioned to me how surprisingly thin the atmosphere appeared to them from space. More general awareness of the importance and scarcity of habitable atmosphere might moti-vate people around the world to care more about how human activi-ties impact the atmosphere.

GOING LOCAL: A TOUR OF THE IMMEDIATE NEIGHBORHOOD

The Earth is part of a star system in the Milky Way galaxy. As most school kids know, it includes one star at the center, orbited by eight planets. Our solar system is not only one star and some planets, of course, but also dwarf planets, hundreds of moons, and comets, plus many thousands of asteroids and meteoroids. The eight planets of the solar system are, in order from closest to the Sun: Mercury,

Venus, Earth, Mars, Jupiter, Saturn, Uranus, and Neptune. The best way to think of them is in two categories. The terrestrial planets (Mercury, Venus, Earth, Mars) are made of rock and metals and have a solid surface. They are closest to the Sun. The outer four planets, known as gas giants (Jupiter, Saturn, Uranus, and Neptune), are almost entirely gas with no hard crust.

The number of planets in the solar system had been nine since the discovery of Pluto in 1930, but discoveries of more icy bodies out in the Kuiper Belt beyond Neptune forced a rethink. Astronomers had to decide: Reclassify Pluto as a non-planet or, for the sake of consistency, be prepared to keep naming more planets indefinitely as potentially hundreds of Pluto-sized bodies may be discovered out there. Pluto lost. In 2006, the International Astronomical Union (IAU) voted to demote the small frozen world to "dwarf planet" status.[349] It's still there, of course, one of our neighbors as it always was. And it can continue to be the target of scientific research and public interest, but just no longer as an official planet. Neil DeGrasse Tyson, an astrophysicist and popular science advocate, was against Pluto retaining planetary status and took a lot of heat from the public for it. But Tyson is at peace with the IAU's decision. "Pluto, I think, is happier out in the Kuiper Belt where it belongs," he writes. "To consider it a red-blooded planet overlooks its fundamental properties. If you were to move Pluto to where Earth is now, it would grow a tail just like a comet, and that is certainly no kind of behavior for a planet.[350]

WHAT DOES IT TAKE TO BE A PLANET?

When the International Astronomical Union (IAU) struck Pluto from the official planetary roll in 2006 the decision was accompanied by the IAU's adoption of the following requirements for earning planethood:

(1) A planet is a celestial body that
    (a) is in orbit around the Sun,
    (b) has sufficient mass for its self-gravity to overcome rigid body forces so that it assumes a hydrostatic equilibrium (nearly round) shape, and
    (c) has cleared the neighborhood around its orbit.[351]

Pluto meets the first two criteria but fails on the last one. It has not gravitationally eliminated nearby objects by pulling them in and absorbing them. With less mass than even tiny Mercury, Pluto lacks the kind of gravitational power necessary to have "cleared the neighborhood around its orbit". But Pluto has many advocates, some of whom point out that there are thousands of near-Earth asteroids that Earth and even Jupiter have failed to clear from their immediate neighborhood. Does that mean they aren't planets?[352]

Jupiter is the most massive and therefore gravitationally powerful planet in the solar system. Named after the top Roman god, it is the largest object after the Sun. It is *huge*. All the other planets in the solar system could fit inside of it if Jupiter were hollow. It's more than 1,300 times the volume of Earth.[353] Jupiter is one of the gas giants; its atmosphere is made up primarily of hydrogen and helium. It has a hydrogen ocean wrapped around its core. The colorful swirls that make it so visually distinct are clouds of ammonia and water. The Great Red Spot is a colossal storm that is more than three times the width of the Earth and has been raging for a few centuries of Earth time. Winds on Jupiter can reach 400 miles per hour.[354]

As a gas planet, Jupiter doesn't have a solid, rocky surface like Earth and Mars, so don't hold out much hope for astronauts landing there one day. It would be a challenge for them to explore anyway. Jupiter's gravity is about two-and-a-half times that of Earth. This means a 160-pound person on Earth would weigh 400 pounds on Jupiter.

<BOX>PLANET NINE FROM OUTER SPACE

There is a possibility that the solar system will again have nine planets. No, Pluto isn't likely to be returned to its former glory. California Institute of Technology (Caltech) researchers have found tantalizing clues indicating that there might be a huge, Neptune-sized planet following an unusual and very distant orbit far beyond Pluto. They are calling it "Planet Nine" but are careful to say that it's not been confirmed but only *predicted* by the data.[355] The clues are in the form of an orbital clustering of objects in the Kuiper Belt that seem to be influenced by something planet-sized. If it does exist, its orbit is so far from the Sun that a single year on Planet Nine—one trip around the Sun—would take 10,000 to 20,000 Earth years.[356]

MOONS ON THE RISE

Planets may get most of the public's attention, but moons can be exciting places, too. They are a significant part of the immediate human neighborhood and some are far more interesting than many people probably realize. The last few decades of scientific exploration and analysis have revealed just how amazing they are. Some are volcanically active. Some have water. Titan, one of Saturn's more than fifty moons, even has a thick atmosphere of mostly nitrogen. Some moons are among the best candidates for hosting extraterrestrial life in the solar system.[357]

Thanks to the Cassini space probe's remarkable work, for example, Enceladus, another one of Saturn's moons, is now thought to have a warm ocean with enough heat activity in the form of thermal vents to possibly support life.[358] Cassini discovered "geyser-like jets" spewing water and ice particles up from an ocean beneath the icy crust.[359] The oceans of Jupiter's Ganymede and Europa moons also offer enticing possibilities for life. Europa, for example, has a deep ocean with tides that may contain twice as much water as the Earth has. Europa also has a thin oxygen atmosphere.[360]

A moon is defined as the natural satellite of a planet and our solar system has hundreds of them.[361] There is an arbitrary element to classifying moons because some planets have millions of small rocks orbiting them. These are "natural satellites", too, but no one thinks of them as moons. To date, there is no official size requirement for moon status. As it is with the life, the solar system is under no obligation to cooperate with our desire for neat and orderly classification systems.

Six of the solar system's eight planets have moons. The exceptions are Mercury and Venus. Jupiter seems to have the most moons of all with more than sixty named or awaiting confirmation. The Earth has just one, named simply the "Moon". Humans have visited the Moon nine times, successfully landing and exploring it on foot six times.

LIFE? Many astrobiologists think Enceladus, one of Saturn's moons, is a prime candidate for life. Towering plumes of water vapor can be seen in this image from the Cassini probe. This activity indicates the possible existence of thermal vents in the ocean below the icy crust of Enceladus and, perhaps even, the presence of life. *NASA/JPL/Space Science Institute*

THE ASTEROID MENACE

In addition to the planets and moons, there also are thousands of asteroids and other objects moving around at high speeds throughout the solar system. Asteroids are mostly the rocky leftovers from the early days of the solar system's formation. Chunks of matter that didn't get sucked up into a forming planet or moon, or haven't collided with one since, are still ripping around, at the mercy of their momentum and something else's gravitational pull. Learning about asteroids can get scary because collisions seems inevitable on a long enough timeline—and impacts can be as profoundly influential as they are cataclysmic. In the case of the Earth, for example, asteroid impacts have altered our geology, our atmosphere, and changed the evolutionary course of life.

PLANET KILLER. Vesta, nearly 300 miles wide, is the largest known asteroid in the solar system. Fortunately, it is not a course that threatens Earth. *Credit: NASA/JPL-Caltech/UCAL/MPS/DLR/IDA*

According to NASA, more than half-a-million asteroids currently are known to science.[362] No one doubts that there are many more, however. Most that we know about fall within an orbital path around the Sun in a zone roughly between Mars and Jupiter. This is called the Asteroid Belt. Some of them are big enough to have their own satellites, (or moons), orbiting around them as they soar through space. A few asteroids have been found sharing a mutual orbit around each other as they move around the Sun. What I find most fascinating about asteroids—apart from their potential to cause death and destruction here on Earth—is that most of them have been around since the formation of the solar system. And they don't erode from wind and water because they are in space, where neither exists. So, apart from craters left by smaller objects hitting them, many are 4.6 billion year-old relics from profoundly pivotal times. As such, they make potentially rich targets for research.[363]

LESS THAN ONE-TENTH OF ONE PERCENT

How could the Earth and every living thing on it be *such a small fraction of the solar system's mass as to be almost comparatively nonexistent*? When scaled against the incredible mind-boggling size and contents of the Milky Way galaxy, the claim makes sense, but when placed against the solar system? Earth is one of eight known planets, the fourth largest, and has significantly more mass than hundreds of moons and asteroids. But the reality is that this neighborhood has three gargantuan residents. The Sun is so massive that it alone accounts for nearly all the known mass in the solar system—99.9 percent of it! Jupiter and Saturn account for 90 percent of the remaining one-tenth of one percent of mass. This means the remaining six planets, all the moons and asteroids, plus every living thing on Earth *combined* make up *less than a tenth of one percent* of all the mass in the solar system.[364]

As child I remember learning that we owe Jupiter gratitude for serving as the solar system's great vacuum cleaner. Its great mass and gravitational power sucks up asteroids (big space rocks) and meteoroids (smaller space rocks) that otherwise might strike the Earth. But more recently computer models have suggested that Jupiter's big bully gravity might sometimes alter the course of an asteroid in a way that could send it *toward* the Earth.[365] The largest known asteroid is Vesta with a diameter of about 329 miles (530 kilometers)[366]. It includes a mountain on its surface that is twice the height of Mt. Everest.[367] For comparison, the asteroid that struck Earth 65 million years ago and likely brought about the extinction of most dinosaur species, was "only" six miles wide. An object named Ceres had been the largest asteroid in the solar system but was reclassified as a dwarf planet in 2006. It is still the largest known object in the Asteroid Belt.

RIGHT NOW, IN A GALAXY *NOT* FAR, FAR AWAY

Once thought to be the home of virtually all the mass and matter of the universe, the Milky Way galaxy, our greater neighborhood beyond the solar system, is now known to be just one of many, many others. In total, there may be from 200 billion galaxies in the universe to perhaps as many as two trillion (2×10^{12}). An estimated ninety percent of galaxies have not yet been studied.[368] The Milky Way is a spiral galaxy which means its contents are arranged into a giant, swirling disk. It is about 100,000 light-years wide and a thousand light-years thick. The solar system rotates with the galaxy at a speed of about 500,000 mph or 800,000 kph in a gigantic orbit around the center of the galaxy. Still, even at this great speed, it takes about 225 to 250 million years for the Earth to complete one complete revolution around the Milky Way.[369]

This is an evidence-based artist's rendering of the Milky Way galaxy, home of *Homo sapiens*. There are no photographs of the Milky Way, of course, because no Earth spacecraft has ever traveled beyond our galaxy to be in a position to photograph it. The entire galaxy is structured as a flat, circular disk made up of many billions of stars. The Sun is identified by name in this image and located below the bright center and near the Orion Spur, which falls between the Sagittarius and Perseus arms. *Image by NASA/JPL-Caltech*

The Milky Way galaxy contains a stunning number of stars but no one knows exactly how many. The most common estimates given by experts range from 100 billion to 400 billion.[370] Bottom line: *It's a lot.* What makes this extraordinary number of stars more exciting is that recent discoveries indicate that not some but perhaps *most* stars have planets orbiting them. I can remember in the past saying to friends and anyone else who would listen: "Just imagine the possibilities if even some of the stars out there have a planet or two orbiting around them." Thanks to science, no imagination is necessary these days. In the last couple of decades, NASA has found thousands of exoplanets (planets outside the solar system), with more being discovered on a regular basis. Most of the credit for this goes to NASA's Kepler Telescope, a space observatory. To date, Kepler has to its credit more than two thousand confirmed exoplanets found and nearly five thousand candidates awaiting confirmation.[371] Thirty exoplanets discovered to date are of high interest because they are similar in size to Earth and orbit their star within the habitable zone, also called the Goldilocks zone.[372] Not too cold and not too hot, NASA defines the habitable zone as "the range of distances from a star where liquid water could pool on the surface of an orbiting planet".[373] Astrobiologists, the scientists who study life on Earth and the possibility of life existing elsewhere, think that planets with liquid water make the most likely candidates to host life of some kind. This is based on what we know about Earth life, of course, which is the only reference we have available. It is possible, of course, that there may be other life-forms in the universe that do not require liquid water.

Experts now say that planets orbiting stars is the rule and not the exception. Estimates based on current survey data put the average number of planets around each star at more than one, meaning there could be a minimum of 1,500 planets within fifty light years of Earth.[374] There also is good evidence that Earth-sized terrestrial planets far outnumber giants like Jupiter. Statistical projections based on the data indicate that at least two-thirds of all stars in the galaxy probably have an Earth-sized planet. This means an estimated minimum of ten billion terrestrial planets are probably in the Milky Way galaxy.[375]

Humankind's home galaxy may be large, far bigger than dwarf galaxies, for example. But it is not even close to being the largest galaxy in the universe. Astronomers know of giant elliptical galaxies that are *twenty times more* massive than the Milky Way.[376] The immense number of stars and planets in the Milky Way might mislead some people into thinking that it's a crowded and congested neighborhood. I suppose in the context of the universe it could be viewed that way, but down at the human level, "crowded" doesn't seem like the best description. *Lonely* might be more appropriate because the stars of the Milky Way are separated by such vast distances. The nearest stars to Earth, for example, are about four light-years away. A light-year is a measure of distance, not time. One light-year is how far light can travel in one year (about 9.5 trillion kilometers or 5.9 trillion miles). So, if NASA were to invent the propulsion technology and then build a spaceship capable of travelling at lightspeed (186,000 miles *per second*), it still would take about four years for it to reach the closest star. Four years might seem doable but don't expect this to happen anytime soon, however. Current rocket technology is not close to this speed. The New Horizons probe, one of the fastest space-craft of all time, moved at a speed of only about 36,000 mph or 58,000 kph. This is so far from lightspeed that if the New Horizons space-craft were heading for the nearest star it would take it about 78,000 years to get there.[377] Four light years, it turns out, is the typical distance between stars throughout the galaxy.[378]

THE ULTIMATE HOME

We know our location within the Milky Way galaxy, but where does this place us in the greatest context of all? Where are we in the universe? Unfortunately, nobody knows. The problem is that the size, shape, and dimensions of the universe are still unknowns, perhaps not even knowable. Therefore, it's impossible to place the Milky Way galaxy at a specific point that gives it a meaningful location regarding the entire universe.

What astronomers do know about the bigger picture is that the Milky Way is part of a giant gravitationally bound collection of galaxies called the Local Group. It's something like a galaxy of galaxies with the Milky Way located near the center. The Local Group is approximately one million light years wide, contains more than fifty galaxies, and rotates around a central region. The biggest galaxy in the Local Group is the Andromeda galaxy, with the Milky Way second largest and the Triangulum galaxy third. Finally, the Local Group lies somewhere inside of an even bigger system called the Virgo Supercluster. It's thought to be about *110 million light-years* wide and there are millions of superclusters in the universe.[379] One of them that is known, the Saraswati supercluster, contains about 400 galaxies and is 650 million light-years wide. Astronomers announced its discovery in 2017. The Saraswati supercluster is approximately four billion light-years away from the Earth and is considered to be one of the largest structures in the universe.[380]

This matters. Location is a big part of who we are. The collective human past, present, and future directly connect to *where we are.* "Understanding cosmology can give you both a sense of scale and a sense of wonder," explained astrophysicist Brian Koberlein.[381] "As children everything is magical, but simple. As we become adults we learn of the subtlety of our lives and our mortality. Understanding the Cosmos can help humanity mature as a species, moving us into an age where we understand our species' place in the Cosmos and the potential for its mortality."

Should anyone ever ask you where you live, you can offer the following science-based and somewhat complete answer to that question: You live on top of the land portion of Earth's crust and within a very narrow band of the lower atmosphere. Your home is the third of eight planets orbiting the Sun, a star that is one of perhaps 300 hundred billion in the Milky Way galaxy. You live about midway out in the vast and flat galactic disc that lies near the center of the Local Group, itself a part of the Virgo Supercluster. It is knowledge like this, basic yet unknown to many, that can enrich a life. To know that we live within something so grand, complex, and majestic as the Milky Way can be inspiring.

John Bochanski is an assistant professor of physics at Rider University who studies the formation and evolution of the Milky Way galaxy. I asked him about the motivations and rewards behind his work. "I grew up fascinated with the Universe around us," he said. "As a child, I wanted to be an astronaut. As a teenager, I told my classmates I'd be an astrophysicist. After receiving my Ph.D., I'd wander outside of the telescope dome to stare up at our Milky Way. At any stage of my life, I could look up at the stars in the night sky and feel a sense of wonder." Bochanski continues:

That childhood wonder is only bolstered by learning more about our Universe and our home, the Earth. That the iron in my blood was formed in a star sometime in the last 10 billion years is not only mind-blowing, but it brings the Universe literally into my very being. I tell my students that we are not "in" the Universe, but that we are a part of it, and a special part at that. While we now know from Kepler observations that small, rocky planets like Earth are common, life is a different story. Life, as far as we know, is not common in the Universe. It has found a special home on our planet. There have been multiple serious extinction events on our planet, but life keeps coming back. Life is further protected by a stable climate, with seasonal changes that are small due to our large, massive moon. This only underscores the obligation that we share to serve as stewards of our planet, protecting the environment and reducing the effects of human-based global warming. Science leads to a better understanding of our home in the Universe, and how unique it is in the Cosmos.[382]

Chapter 9

HOW DO BRAINS WORK?

W e live in fortunate times. Thanks to science, many basic things about the human brain have become common knowledge. We know that it is the fountain and fortress of all thoughts and dreams. We understand that the brain is made up of many distinct components with particular functions and responsibilities. Because of science there can be no doubt about the brain's unique importance. It is everything to the individual, the ultimate source of intelligence and personality while also serving as command and control center for critical body functions such as breathing and heartrate. Practical, evidence-based ideas about brain health and performance are available today for anyone to adopt. Perhaps the most amazing, some might say disturbing, thing about the brain is that subconscious or unconscious processes—thoughts and perceptions the host is not aware of—utterly dominate the brain's activities. Most people do not know or appreciate this but virtually all conscious judgements and decisions are unconscious commands in disguise. Some experts are convinced that *all* the brain's output is of an unconscious nature and that our feelings of conscious autonomy are but a delusion. More on that later. Many mysteries remain about the brain but there is today an abundance of knowledge about its many structures and their functions ready and waiting for anyone to learn and make practical use of. This was not always the case, of course. A reality based view of the human brain was a long time coming.

Hippocrates, a physician in ancient Greece, seemed to have the brain somewhat figured out some 2,400 years ago when he wrote: "Men ought to know that from nothing else but the brain come joy, despondency and lamentation."[383] But it would take many centuries before most people on Earth were aware of even the most basic facts relating to the brain. The ancient Egyptians, for all their technical brilliance and extreme reverence for the dead, seemed to have unceremoniously yanked out the brains of the deceased via the nostrils during the mummification process and tossed them out with the trash. No human brain has ever been found in an ancient Egyptian tomb and there are no records of them ever being specially preserved.[384] Aristotle may have had a great brain but he didn't seem to know much about that part of his anatomy. He thought the brain was a secondary organ, a mere radiator that circulated blood to help cool the heart.[385] The brilliant ancient Roman physician Galen managed to deduce, based on his observations of people with brain injuries, that thinking took place in the brain. But even this basic fact would not become universal knowledge until many centuries later. Galen also thought that the brain housed the "animal soul", one of three souls in the body, and that the brain was cold and made of sperm. He was not an outlier regarding that latter belief. It was widely believed in Europe and Asia that the brain was either made of sperm or existed primarily for the purpose of making sperm. Even the Renaissance genius Leonardo da Vinci, who completed very complex and detailed studies of the brain, seemed to think this to some degree, having sketched what he described as a semen-delivery tube that ran from the brain to the penis.[386]

The brain has never completely escaped many of the past misconceptions about it. It is common now, for example, to speak figuratively of the heart as the seat of love or determination. While that may be a harmless figurative holdover, some misconceptions are dangerous. There are people today, for example, from London to Laos, who continue to cite demonic possession as an explanation of some mental illnesses or unapproved behaviors.[387] Knowing something about the evolutionary history, structure and functions of a human brain is no trivial pursuit. This is a fundamental key to improving individual lives and entire societies. The better we know the brain the better we know ourselves. There is no doubt that we can make better use of this three-pound blob of electrochemical magic that evolution has gifted every human. For example, recent research has found that, in some contexts, rational thinking can be improved relatively quick and easy.[388] Given the crucial role of the brain in every life, it is remarkable that so few people take the time to learn much if anything about it. It seems to me negligent and irresponsible that every elementary or primary school does not provide all students with an extensive education on the structures, workings, and requirements of the brain. This is essential knowledge.

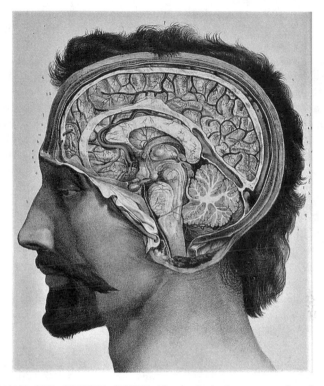

COMMAND AND CONTROL CENTER. The brain is the source of endless conscious and unconscious thoughts, emotions, creativity, learning, and memory. It also regulates many essential body movements and functions such as walking, sleeping, heartrate and breathing. The brain is the most important organ, by far, which makes it all the more strange that most people know little or nothing about its structures and processes. *Illustration from* The Atlas of Human Anatomy and Surgery, a 19th century French textbook *by J. M. Bourgery and N. H. Jacob*

We know that the human brain has immense powers and abilities. But it's important we not go so far in praising and admiring the brain that we miss its many challenges and problems. Yes, special wonderful and productive as it is, this is also one confounding mess of a thinking system. Every brain's standard systems and natural functions consistently mislead and make fools of its host. And most people have no idea how any of it happens or what can be done about it. The glorious brain that has produced so much genius and beauty is far from a well-designed, reliable perception and computation device. Even a cursory study of the brain reveals that it is not a perfect design. Evolution patched it together on the run and this shows. It's understandable, of course, because there are no time outs in the natural world. It's not like nature could have placed Australopithecines and *Homo erectus* in suspended animation or dry dock for sensible intracranial construction. Life forms have to keep moving, living, and surviving while evolving. New York University psychology professor Gary Marcus calls the modern human brain a "kluge". This is an engineering term for a workaround or clumsy by effective solution to a problem. "Nature is prone to making kluges," Marcus writes in his book *Kluge: The Haphazard Construction of the Human Mind*, "because it doesn't 'care' whether its products are perfect or elegant. If something works, it spreads. If it doesn't work, it dies out. Genes that lead to successful outcomes tend to propagate; genes that produce creatures that can't cut it tend to fade away; all else is metaphor. Adequacy, not beauty, is the name of the game."[389] This is the brain's way: learn, make do, get by, and figure it out. Every human brain is a dense maze of neural shortcuts, organic algorithms like galaxies that are always reconfiguring themselves. Those that work well enough and often enough earn their keep.

Humans are born into the world helpless and unprepared and then stay that way longer than any other animal. Wildebeests calves are up and active, able to run for their lives, within less than an hour after being born. I have seen this and it's remarkable how fast it happens. The wildebeest mother goes to work, within minutes of the birth, pushing her baby, forcing it to walk and run so the nervous system and muscles will be ready for action should a predator show up. Dolphins can swim immediately after birth. For *Homo sapiens*, however, years of dependency are required before some minimal level of competence and independence is attained. Human babies lie around for years depending on their parents for everything. I currently have two teenagers living in my house who still seem to lie around and depend on their parents for everything. It takes human so long to get up to speed because people are *brain creatures.* Although unconscious instinct, shortcuts, hunches, and emotions are prominent and constantly flowing up, humans rely on acquired conscious knowledge to a very high degree, as well, much more than any other species. This is because long ago the nonrandom forces of evolution nudged our genus onto and then far down a path that became increasingly committed to cognition-based survival. What our ancestors lacked in sprint speed, strength, teeth/horn/claw weaponry, and size, the brain would make up for. The key is learning, always learning. Babies must learn. Children must learn. And adults continue learning.

Flexibility and creative problem solving were beneficial to human ancestors so those traits were selected for across millions of years, resulting in a modern brain that can imagine almost anything and productively grapple with many incredibly difficult challenges—all while simultaneously monitoring and controlling vital life functions throughout the body. The brain is the reason prehistoric humans were able to spread to so much of the planet on foot and settle in so many diverse lands. It is the reason that the humans of modern history have been able to achieve so many remarkable things. The brain can be a source of endless wonder and immense pride for everyone. It was the single most important key to success and survival against predators and environmental challenges that could have wiped out the hominin line many times over the last few million years. The brain-based intellectual teamwork we call human culture took us from fearing the night to owning it with fire and weapons. But the brain's greatness is not obvious beyond the results it produces.

Like many experts, Ian Tattersall, curator emeritus of the American Museum of Natural History in New York City, views the human brain as nothing that screams extraordinary gifts and abilities. It is mostly a typical ape brain, scaled up and turbocharged, perhaps, but not so special otherwise. And yet somehow it ended up with profoundly special abilities. "The human brain, whatever its marvels, probably does not contain any completely new structures—any structures, indeed, that are not shared with all of our primate—or even mammal—relatives, however humble," he writes in his book, *Becoming Human: Evolution and Human Uniqueness.*[390] "Thus we cannot look merely to entirely novel brain components to explain our cognitive powers, however elegant an explanation that would be. What has happened over our evolutionary history, however, is that certain parts of the human brain have become enlarged or reduced relative to others and the connections between them modified or enhanced. Even this is not unique to us, though: for while we undeniably have the largest primate cerebral cortexes (about 76 percent of our large brain's total weight), there has been a dramatic increase in the percentage of the brain occupied by the cerebral cortex and supporting structures among higher primates in general."[391]

A typical human brain is about the size of a small cantaloupe. It measures about six or seven inches from front to back and has a volume of around 1100 to 1200 cubic centimeters. A typical adult brain runs on only about ten to twenty watts of electrical power, about what is required to dimly light a small light bulb.[392] An adult female brain weighs approximately two and a half pounds and male brains around three pounds. The trend in human evolution was toward larger brains. But the correlation between size and intelligence is weak. Whale and elephant brains are much larger, for example, and yet whales and elephants can't do calculus. Neanderthal brains were significantly bigger than modern human brains.[393] Apparently, however, those couple of hundred or so additional cubic centimeters of volume couldn't help them to avoid extinction. African elephant brains are about three times larger than human brains and contain 257 billion neurons.[394] This is about *triple* the human brain's 86 billion neurons.[395] At 1,230 grams Albert Einstein's brain was slightly below average for a modern male brain.[396] So, while size does equate to brain power in a general sense—a chimp's brain is larger and has more intellectual potential than a fly's brain, for example—absolute size is far from everything.

The ratio of body mass to brain size also fails to explain the extraordinary intelligence of people compared with other animals. The human brain accounts for about two percent of body mass while the tree shrew's brain is some ten percent of the animal's overall mass.[397] Some very small species of ants have brains that are 15 percent of body mass.[398] Imagine a human with a brain that is 15 percent of body mass. The head would be more than seven times larger than normal and the brain would weigh about twenty pounds instead of three. What seems to matter most when it comes to intelligence is not how big the brain or how many neurons it contains but *how* the brain is put together. Modern humans, for example, have a unique abundance of neurons in the cerebral cortex, the part of the brain crucial to complex analysis and higher reasoning. Another key to the human brain's astonishing capabilities is the *way* in which neurons connect, share, store and retrieve information throughout life. The *quality* of their many networks determine intelligence. Human intelligence, then, is ultimately a function of *connections*.

Elaborate and intricate networks fill the human brain to make it the unprecedented engine of problem-solving and infinite creativity that it is. I believe it may be more useful to think of the human brain as less of a thing or an organ and more of a *process* or *performance*. In the way that a great play is something more than a script, a director, and a bunch of actors, the brain is substantially more than cells and fluid sharing space inside a human head. Its more a musical riff that comes from the piano than the piano itself. A human brain constantly rewires itself throughout life, for better or worse. Therefore, it is forever a work in progress until death finally silences it. Every feeling, thought, memory, and new thing learned joins the galaxies within and makes them anew. "There are billions of neurons in our brains, but what are neurons? Just cells. The brain has no knowledge until connections are made between neurons. All that we know, all that we are, comes from the way our neurons are connected."[399] That's from Tim Berners-Lee, the inventor of the World Wide Web and a man who knows a thing or two about networks and connections.

WHAT'S ON THE BRAIN'S MENU?

Modern scientific research has revealed a lot of invaluable information about what is best to eat and what should be avoided to nurture and protect our brains over a lifetime. For example, excessive consumption of *refined sugar may be harmful* to the brain. Many scientists now suspect that high-sugar diets are linked to greater risk of dementia and depression.[400] Brain derived neurotrophic factor (BDNF) is a vital chemical substance in the brain that is required for learning and memory and high sugar consumption, research suggests, *reduces the amount of BDNF in the brain.*[401] One recent study, published in the journal *Neurology,* found that people who consumed excessive amounts of refined sugar—this included people who were currently healthy and nondiabetic—experienced degraded memory and a general "negative influence on cognition [thinking], possibly mediated by structural changes in learning-relevant brain areas."[402] In another experiment, researchers gave sugary drinks to rats and found that the rats suffered losses in memory and learning abilities. Adolescent rats suffered the worst effects.[403]

Both the body and brain need Omega-3s (polyunsaturated acids) for many regular functions. But we can't produce them ourselves, so we depend on plants and animals to get them. Omega-3 sources include *salmon, kale, spinach, flaxseed, walnuts, and kiwi fruit.* It might be helpful to think of your food in general as a "pharmaceutical compound that affects the brain".[404] Being aware of the impact food choices have on brain health and performance can be powerful motivation to eat well.

Leafy green vegetables are one of the best foods for the brain, according to the credible research I have reviewed. One study published in 2015 tracked the diet and mental capabilities of nearly a thousand older men and women. To focus in on the impact of diet, the scientists controlled for education, gender, smoking, physical activity, and genetic risk for Alzheimer's disease. Remarkably, those who ate one or two daily servings of green leafy vegetables suffered significantly less cognitive decline than those who did not. The spinach and kale eaters had the mental fitness typical of people *eleven years younger.*"[405] Vitamin K, beta-carotene, lutein, and folate, and were the specific nutrients in the vegetables believed to have aided the subject's brains.

Blueberries, blackberries and strawberries contain flavonoids called anthocyanins that have been shown to improve the mental performance of animals. There is strong research showing that these berries can improve human cognition as well and may even help to prevent loss of memory and other age-related brain problems.[406] Researchers also have found a link between monounsaturated fatty acids and general intelligence. These fatty acids are found in *walnuts, olive oil, and avocados.*[407]

Anyone who hates avocados and spinach, or just wants to consume these nutrients fast and conveniently, can try doing what I do. Four or five times per week I make my own personal shake based on the best current scientific, evidence-based recipe I can come up with. And I love it. It tastes great while also giving me the nice feeling of doing something helpful for my brain—the one irreplaceable organ that sits at the center of my personal universe. I simply toss the following ingredients into a blender (not a juicer) and mix them up: *spinach and/or kale; blueberries and/or blackberries; a banana; a carrot; an avocado, a few walnuts; a teaspoon of flaxseed; ice cubes and water.* I suggest adjusting the ratios to suit individual taste. I used to require a bunch of strawberries for the shake to taste good to me. Over the years, however, I've been able to reduce the strawberries (less sugar) and increase the amount of spinach and kale, without sacrificing any of the taste or pleasure I derive from drinking the shake. To be clear, I am suffer no delusions that this shake is the magic elixir makes me smarter, will keep me live until I'm 120 years old, or keep me cancer free. I just think it makes sense to consume it because science has shown that the ingredients are good for human bodies and brains.

THINKING ABOUT THE CELLS THAT THINK

Achieving a basic understanding of neurons, the specialized brain cells that make thinking possible, is fundamental to knowing the brain. In every person there is a dense and tangled neural thicket. Countless branches sprout from tens of billions of neurons. To imagine a glimpse of the brain's internal workings is to see a bizarre but awe-inspiring labyrinth of thoughts in motion. A typical brain contains an estimated 86 billion neurons, a galaxy of stars. But it's even more than that because all of these neurons are neither independent nor stagnant. They move to connect with one another and share information, to form elaborate networks. They break free to modify existing networks and create new ones, numbering in the tens of millions. The numbers of connection all this relentless activity creates is stunning: perhaps *a hundred trillion connections* at any given moment.[408] There are many more neural connections in a single human brain than there are stars in the Milky Way galaxy. And, remember, these networks are in a constant state of flux. New experiences, internal interactions with old memories, fantasies, and so on keep the brain's landscape in motion. Every person begins each day with a new brain, one that is physically different from yesterday.

"When we learn something, whether it's a French word or a salsa step, cells morph in order to encode that information," explains John J. Ratey, associate clinical professor of psychiatry at Harvard Medical School, "the memory physically becomes part of the brain."[409] This changing condition of the brain sends its complexity through the roof. The prominent neuroscientist V. S. Ramachandran claims there are many more possible network variations within a single brain than there are elementary particles in the universe.[410] Elementary particles are parts of atoms, so he is claiming that there are many more network possibilities, or potential states, within a single brain than there are atoms in the entire universe. This may seem absurd at first glance given the number of atoms it takes to make trillions of stars, planets, moons, etc. It begins to make sense, however, when you consider that each one of the 86 billion or so neurons in a brain can make connections with up to ten thousand other neurons during its individual lifetime.[411] The number of potential combinations is so large as to be incomprehensible. There is a universe stirring within every human brain.

Neurons have the standard cell parts: nucleus, DNA, cytoplasm, mitochondrion, vacuoles, and ribosomes. But they look very different from other human cells and go about their duties in unique ways. Resembling a tentacled sea creature or maybe some classic sci-fi monster, a neuron connects to another via the dendrites (the "tentacles") of one and the axon ("tail") of the other. Without physically touching they draw extremely close and then transfer electrical-chemical impulses from one to the other. These electrical impulses exit the neuron at the axon and impulses are received through the dendrites. Fatty tissue called the myelin sheath is wrapped around the axon to serve as an insulator for speeding up the signals. The point of union between an axon and dendrite is called a synapse. When one of the tiny electrical charges surges into a neuron it triggers the release of a tiny batch of chemicals. These are called neurotransmitters. Amazingly, these miniscule bits of information "swim" across the narrow gap from the axon and into the dendrite.

Once the information is transferred, the neuron that received it has the option of holding it there or sharing it with another neuron. This is how every thought in your brain is created, processed, and stored, from the memories of your favorite childhood toy to how you imagine you might fare in a zombie apocalypse. Right about now your brain may be eagerly anticipating reading an explanation of how tiny jolts of electricity and chemicals zap around between billions of neurons to make thoughts and memories. Unfortunately, no such explanation is available. Not yet. Enough is known now, however, for us to use toward enhancing our brains.

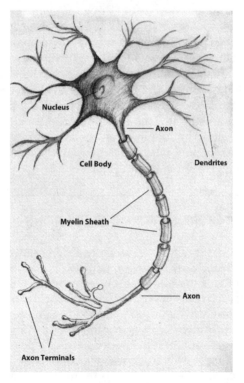

These highly specialized brain cells are the currency of thought. A typical human brain has some 86 billion of them. Neurons link up to create intricate networks that are basis of reasoning, creativity, and memory. Science recently revealed that we can generate new neurons throughout life by learning new skills and consistently engaging in physical activity. *Illustration by Sheree Harrison*

It can be exciting, even liberating, to learn that one is not born with a static, frozen brain. No one ever need be fatalistic or feel passive about the condition or performance of his or her brain, not according to science. Genetics, injury and disease cannot be brushed aside, of course, but significant positive enhancements can be made to any brain through dedicated, smart effort. Ratey continues: "What we now know is that the brain is flexible, or plastic in the parlance of neuroscientists—more Play-Doh than porcelain. It is an adaptable organ that can be modeled by input in much the same way as a muscle can be sculpted by lifting barbells. The more you use it, the stronger and more flexible it becomes."[412]

One of the most important things anyone can learn and remember about the brain is that it is capable of *growing new neurons* throughout life through learning new skills and by consistent physical activity.[413] It was once thought that a person was born with a finite number of neurons and that was that. Now we know better. You belong to one of the first generations of people who have access to this profound and useful knowledge. Brain science has given all of us powerful new motivation to read, write, sing, dance, hike, run, swim, play challenging games, paint, sculpt, create music, do math, eat wisely, and get sufficient sleep because these things are critically important for our brains.

MORE SWEAT MORE NEURONS

Physical activity literally generates new brain cells. Regular and sustained aerobic activity—say, a daily walk, run, bike ride, or swim—stimulates the hippocampus to produce more neurons in a process scientists call *neurogenesis*.[414] It is in everyone's interest to have the hippocampi as healthy and productive as possible because they are key centers for learning, memory, and mood regulation. Happy hippocampus, happy life.

The evidence for the exercise-brain connection is powerful. For example, one large and long-term study tracked 1,500 people in Finland across twenty-one years of their lives and discovered that those who had exercised at least twice a week on average were *50 percent less likely to develop dementia* than the people who in the study who did not exercise regularly.[415]

Exercise increases blood circulation in the brain and this helps keep the important glial cells healthy.[416] Exercise is also known to boost levels of BDNF in the brain, the crucial chemical substance involved with learning and memory.[417]

Physical activity also can improve moods and even alleviate the pain of depression in some cases. Research has shown that *consistent aerobic activity for some people can be more effective in fighting depression than prescribed antidepressants.*[418] Psychiatry professor John J. Ratey describes the value of physical activity in his book, *Spark: The Revolutionary New Science of Exercise and the Brain*:

> The beauty of exercise is that it attacks the problem from both directions at the same time. It gets us moving, naturally, which stimulates the brain stem and gives us more energy, passion, interest, and motivation. We feel more vigorous. From above, in the prefrontal cortex, exercise shifts our self-concept by adjusting . . . serotonin, dopamine, norepinephrine, BDNF, VEGF, and so on. And, unlike many antidepressants, exercise doesn't selectively influence anything—it adjusts the chemistry of the entire brain to restore normal signaling. It frees up the prefrontal cortex so we can remember the good things and break out of the pessimistic patterns of depression. It also serves as proof that we can take the initiative to change something. This paradigm holds true for exercise's effect on mood in general, regardless of whether we're depressed or coping with some nagging symptoms. Or even if we're just having a bad day."[419]

Don't be discouraged if you happen to be living a sedentary life and find thoughts of intense exercise to be terrifying. *Some exercise is better than none.* Science shows that even minimal, moderate physical activity—as little as twenty or thirty minutes, just three times per week—can have significant positive effects.

GEOGRAPHY OF THE BRAIN

The anatomy and the many processes of the human brain are not easy topics to master. But why would they be? The brain is *the most complex thing in the known universe.* There is much about it that even the best experts today do not understand. But no one should be intimidated. Fortunately, even knowing a little bit about the brain can go a long way. The first step is to get the lay of the land and learn the basic parts, and what each is known for doing. This simple step can significantly raise one's appreciation for the unseen brain that so much depends on. Knowing what the brain looks like in some detail, and how its put together, might make it feel more real and relevant to the owner's life. Just this basic awareness alone could perhaps inspire more thoughtfulness about how to use it well and care for it, too.

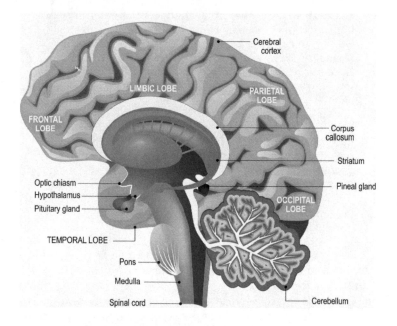

Different people describe the brain in different ways. This can be confusing when a layperson tries to learn about it. With this in mind, I will attempt to present a sensible and straightforward presentation that avoids *overcomplicating this most complex thing in the known universe*. We will survey just four general parts or regions of the brain. They are: *the brain stem, the limbic system, the cerebellum*, and *the cerebrum*. It is important to remember, however, that the brain is best thought of as one unit because it all works together. Memory, for example, involves many areas throughout the brain. So no one should imagine that these four parts or regions operate with autonomy in isolation. I think of the brain as an ecosystem, like a rainforest or coral reef, with many parts connecting to create a larger and more complex sum. Finally, keep in mind that the information to follow is about you. This is personal. You have one of these amazing organs in your head and it is hard at work right now. Think about it often as you read the next few pages. Think of it often for the rest of your life.

BRAIN STEM

Located at the bottom of the brain, the brain stem is the critical connection between the brain and the rest of the body. The brain stem and the thalamus that sits on top of it may not get as much glory as other parts of the brain, but know this: you wouldn't last a minute without them. Collectively the stem and thalamus control many key functions, including: hunger, thirst, body temperature, sleeping, breathing, and blood pressure.

The stem portion looks like it could be a vegetable stalk of some kind. It extends from inside the cranium down through a round opening at the bottom of the skull called the foramen magnum where it connects the brain to the spinal cord. The thalamus is located above the stem portion. This is the relay station, more like Grand Central Station, really. The thalamus processes and directs incredibly large amounts of information as it comes and goes at lightning speeds to and from the brain and all points below.

The brain stem is the reason you keep breathing and don't die should you faint or get struck on the head and lose consciousness. Just imagine if breathing and heartrate had to be supervised, consciously, by another part of the brain. Think of the burden, the toll, the distraction, of having constantly to think to inhale and exhale, to flex your heart every second or so in order to live. Forget either and you die. How would anyone ever sleep?

THE BRAIN THAT TIME FORGOT

Many experts point out that the modern human brain is a poor fit for the modern world we have constructed. Having evolved over millions of years in the prehistoric environment, many of the brain's structures, processes, and unconscious short-cuts that may have been necessary for survival then are less effective now and often the cause of serious problems. Dean Buonomano, a leading neurobiologist at UCLA's Brain Research Institute is one who holds this view and writes about it in his book, *Brain Bugs: How the Brain's Flaws Shape Our Lives*:

> Although we currently inhabit a time and place we were not programmed to live in, the set of instructions written down in our DNA on how to build a brain are the same as they were 100,000 years ago. Which raises the question, to what extent is the neural operating system established by evolution well tuned for the digital, predator-free, sugar-abundant, special-effects-filled, antibiotic-laden, media-saturated, densely populated world we have managed to build for ourselves? . . . Our feeble numerical skills and distorted sense of time contribute to our propensity to make ill-advised personal financial decisions, and to poor health and environmental policies. Our innate propensity to fear those different from us clouds our judgment and influences not only who we vote for but whether we go to war. Our seemingly inherent predisposition to engage in supernatural beliefs often overrides the more rationally inclined parts of the brain, sometimes with tragic results.[420]

Hank Davis, a professor of psychology at the University of Guelph in Canada, warns us about blindly relying on what are essentially prehistoric cognitive operating systems. "In some ways, we are the lucky ones," he writes in his book *Cave Man Logic: The Persistence of Primitive Thinking in a Modern World*. "Our civilization has never been as knowledgeable as it is today. There has never been as much scientific understanding of the world around us as there is right at this moment. . . . Yet, because our minds are no more evolved than they were a thousand years ago, we are just as vulnerable to illusion, comfort, and social pressure as our ancestors were. Faced with incomplete sensory evidence, we are just as likely to come to the wrong conclusion and then fight passionately to maintain those views because of the comfort or stability they provide. We may be less ignorant than our ancestors, but all the hard won additional information does us no good if we don't let it inform our thinking."

One of the practical lessons that come from learning about the human brain is that it needs some help. We must use our brain to help our brain. As powerful, wonderful, and productive as it is, this is an organ caught out of time, shoved unprepared into the crowded, loud, fast, toxic, and stimulating stew of modern civilization. Without time to evolve, the brain can only call on structures and processes better suited for a long-lost past. Constructive skepticism, critical thinking skills, and the scientific method—applied often and vigorously—aid the brain in keeping its host safe, prosperous, and alive.

THE LIMBIC SYSTEM

The limbic system is a fascinating collection of structures that lie above the brain stem in the middle zone of the brain. Think of it as the "feeling brain" because this is ground zero for the most intense emotions. Fear, terror, rage, and sexual excitement well up from the limbic system. Let's take a look at two of the most important components of the limbic system, the Amygdala and the Hippocampus.

Every normal brain includes two Amygdalae. One is located about four inches directly behind the rear side of each eye. They are small, about the same size and dimensions of an almond. But don't underestimate the important role they play. They receive sensory input—a sound, an image, a smell or touch—and if it is determined to be associated with a threat the Amygdalae launches immediate activation of fight-or-flight measures throughout the body. This includes dilation of the pupils, faster heartrate, and a jump in metabolism. Anything can trigger this, from hearing a lion's growl on a dark night out in the African bush, to the sight of masked man with a knife running toward you in a vacant parking lot, or feeling what might be spider legs walking on the back of your neck. The Amygdala have obvious survival value for humans because speed is crucial in many life or death situations.

The Amygdalae seem to have larger implications beyond making our hearts pound in dangerous situations. Two interesting studies published in 2014 found significant implications tied to the size of Amygdalae. One study discovered that men with smaller-volume Amygdalae had higher than expected rates of "severe and persistent aggression and the development of psychopathic personality."[421] The other study revealed that people who had donated a kidney to a stranger, an act of extreme altruism, had Amygdala that were bigger and more active than people in a control group.[422]

The Amygdalae are not only about fear and rage, however. If the senses deliver input that is interpreted as something exceptionally positive, then the Amygdalae respond by flooding the host with pleasurable sensations. In this way they steer the body away from dangers and toward things perceived to be good for us or at least highly enjoyable. They have a long memory, too. If something terrible happens, say nearly fatal encounter with fire in childhood, that person's Amygdala likely would encode the experience and file it away. Then, even many years later, the mere smell of smoke might activate the fear response in an instant, even before the person consciously recognized the smell.

THE PERILS OF A SLEEPY BRAIN

The research is loud and clear about the value of sleep to the brain. Very important things occur in the brain during sleep. Waste and debris are cleared, memories are rehashed and, if deemed potentially valuable to future needs, laid down as long-term memories. A sleep-deprived brain can suffer degraded thinking abilities, reduced memory, and may be more prone to accidents and even unethical behavior.[423] Insufficient sleep is linked to *higher risks for diabetes, heart disease, cancer, strokes, Alzheimer's disease, and obesity.*[424] Failing to get enough sleep also can have a huge impact on mood the next day, even more than income or marriage problems.[425]

Neurologist Paul Bendheim on the importance of sleep: "We often think of sleep as not doing anything, but . . . while you sleep, parts of your brain are hard at work building stronger, lasting memories and practicing tasks you learned the day before. Sleep improves all other critical brain functions—learning, thinking, planning and executing complex goals, creativity and problem solving. Restful, restorative sleep is essential for overall health. You will live longer and in better brain and body health if you pay attention to sleep."[426]

Deep sleep is critical to the brain's *learning efficiency* before and after. If a person does not get enough deep sleep, known as REM sleep, the night *before* then learning ability is likely to be compromised or less than optimal. If a person does not get adequate deep sleep the night *after* learning something, it is likely that the knowledge or skill will not be retained or learned as well as it might have been with better sleep.[427]

The hippocampus is a tiny structure in the brain, about an inch or three centimeters wide, that is key to spatial awareness, i.e., knowing and remembering location. If you find your way home tomorrow, thank the two hippocampi in your brain. The name is Greek, Hippo for "horse" and campus for "sea monster". Early researchers thought it resembled a seahorse. It seems redundant or unnecessary to say that the hippocampus is important because everything in the brain is important. However, life without healthy and fully functional hippocampi would be difficult if not impossible because they sort and process experiences into memories. They determine what becomes a short-term memory for quick use and what gets sent off for storage as a long-term memory for possible retrieval much later. Memories are the basis of learning, relationships, planning, and so on. Therefore, the work of the hippocampi is invaluable. As with the Amygdala, scientists have found that size is important. Generally, the larger the hippocampus, the better. Smaller hippocampi or those that are in the process of shrinking have been linked to Alzheimer's disease and memory loss problems.[428] Fortunately, there is some essential science to enhance your life regarding the hippocampus. You can *grow* it larger through consistent learning and physical activity throughout your lifetime.[429] It's amazing but true. Researchers have found that reading books, creating art, learning new skills, and engaging in regular exercise can not only reduce the deterioration of an aging hippocampus, those things might make it grow larger and stay healthier longer, too.

HOW MANY NEURONS?

The vast majority of books and documentaries about the human brain cite the number of neurons to be one hundred billion. While writing a previous book, I searched for the source of this claim and discovered that apparently none exists. It does not seem to be supported by evidence. The 100-billion-neuron claim seems to be nothing more than an agreed-upon estimation.

Fortunately, a team of scientists decided to find out just how many neurons a typical human brain really does contain. They couldn't just count them one-by-one, of course, because that would take centuries. So they liquified brains in blenders—one at a time—made sure the neurons were equally distributed in the resulting fluid, and then counted sample portions. This allowed them to extrapolate a total neuron count for the entire brain. The total number of neurons for a human brain, the researchers came up with was eighty-six billion.[430] This is significantly lower than one hundred billion but still an astonishing amount. The surface area of this many neurons has been estimated to be the equivalent of more than three football fields.[431]

THE CEREBELLUM

The cerebellum, Latin for "tiny brain", is a marvelous, fascinating, plum-sized component of the human brain. Stuck about halfway into the rear, lower portion of the brain, the cerebellum has a distinctive appearance. It's easy to spot in pictures of the brain with its deep horizontal grooves and darker color. The cerebellum is highly convoluted. (Convolutions are the wrinkling seen all over the surface of the human brain, is a way of cramming a lot of surface area into a limited space.) The cerebellum is so convoluted that it would span more than 1,100 square centimeters if flattened out.[432] That's roughly the equivalent of two standard-size sheets of paper.

The cerebellum is also extremely dense with neurons, packing more than all the rest of the brain combined with about seventy percent of the total neuron count.[433] So what does it do with all those brain cells? It coordinates balance, posture, and physical movements, especially fine motor skills like typing or playing a musical instrument. But it also lights up like a Christmas tree during brain-imaging tests in association with more routine activities such as hearing, smelling, and pain perception. One of the most amazing things about the cerebellum is that all those neurons it has are wired together in much the same way they have been through some *four hundred million years* of vertebrate evolution.[434] In the cerebellum we can see human kinship with alligators, dolphins, and horses.

Neurophysiologist Jim Bower thinks the cerebellum needs more attention. "It is a strange and disturbing structure," he says, "with relatively few people working on it. If you remove the cerebellum in young humans—or animals—[it] turns out there is very little noticeable effect a few months later, [which is] rather odd given how many neurons are there and how much energy the cerebellum consumes. But this [mystery] mostly speaks to how primitive our understanding of human behavior is."[435]

GLIAL CELLS MATTER, TOO

Neurons are never alone. Not only do they have each other, numbering in the billions, they also are surrounded by glial cells. These critical cells help neurons by nudging them along to make their connections, and assist in keeping them in place once they do. They can be thought of as the blue-collar work crew responsible for keeping the celebrity neurons ready for action. Their exact numbers are unknown. It was once widely thought that glial cells outnumbered neurons ten to one, but today some scientists think their numbers may be roughly even.[436]

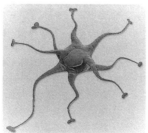 Glial cells may be overshadowed by the glamorous neurons but they perform many essential maintenance and construction tasks in the brain. *Image by Medical gallery of Blausen Medical 2014*

Glial cells deserve more attention from researchers and the public (most people have never heard of them) because in addition to their routine chores like supplying neurons, disposing of dead ones, and clearing waste and debris from the brain, they also help to defend the brain against invading germs. They have been connected to some brain diseases, including Alzheimer's as well.[437] Glial cells may even have an important role in memory formation, too.[438]

THE CEREBRUM

The cerebrum and its uppermost surface area, called the neocortex or cerebral cortex, dominate the brain's overall mass. When most people think of the human brain they probably visualize the cerebrum more than anything. This upper portion of the brain stands out most in models, illustrations, and photos of the brain. Its surface is highly convoluted, with numerous ridges and valleys providing the necessary space for tens of billions of neurons. The cerebrum is the focus of most casual attention and serious research because this is where higher thinking, creativity, and problem-solving take place.

Without a cerebrum, we all would be fish-brained bipeds with opposable thumbs and no idea what to do with them. Think of all that has come from this wrinkled lump of pink tissue. Every painting and every sculpture began in the cerebrum. Every cure for disease, every religion, every war, and every math equation germinated here. *Homo sapiens* may have achieved spaceflight, built weapons powerful enough to destroy civilization, and launched space telescopes peer back through billions of years in time, but no computer to date can match the overall power, complexity, flexibility, and creativity of a human brain.

The neocortex is sectioned off into four lobes, each one with its respective duties. They are:

- **The frontal lobe** (front of cerebrum) is responsible for complex problem solving, making decisions, and planning.
- **The parietal lobe** (top, just behind frontal lobe) helps process sensory information such as heat, cold, pain, pleasure, bright, dark.

- **The temporal lobe** (sides) is involved with emotions, hearing, memory, and language.
- **The occipital lobe** (rear) helps with vision by processing sensory input from the eyes.[439]

The modern human neocortex is a remarkable, though unintended, result of millions of years of brain evolution. It's incredibly thin at a mere two-tenths of an inch thick (less than half a centimeter) but so much of it is crinkled up inside the cranium that it makes up about three-fourths of the brain's total weight. Flattened and spread out, the neocortex would span an area just a bit over seventeen square feet (1.6 square meters).[440] That's about the equivalent surface area of a twin bed.

NO ONE IS EVER FULLY AWAKE

One of the most helpful things one can learn and remember about the workings of the human brain is that most of its processes occur somewhere beneath a very thin surface of consciousness. We may believe our conscious mind is doing almost everything, but it turns out that it accounts for very little of the brain's processes and output. The thoughts and actions of *Homo sapiens*, self-named "Thinking Man", are far more dependent upon intuition and unconscious processes than people are generally aware of. We may like or need to view ourselves as rational, awake, and conscious beings, but the reality is far different. Zombies might be too extreme of a metaphor, but humankind could at least be fairly described a species of sleepwalkers, given that *the overwhelming majority of brain activity is of the unconscious variety*. The role of the conscious mind seems to mostly be in the service of the unconscious mind as delivery agent and then explainer or justifier after whatever has been thought, said, or done.

Ezequiel Morsella, associate professor of psychology at San Francisco University, says we are never fully aware of these processes and we do not control them. He has attempted to explain the dominant role of the unconscious mind with his "Passive Frame Theory".[441] It asserts that humans do not have anything near the free will most believe in and are instead almost always following the lead of the unconscious thoughts and reactions. The conscious mind, according to Morsella, is like a language interpreter. "The interpreter presents the information but is not the one making any arguments or acting upon the knowledge that is shared." Morsella said this in a statement released with his paper on the theory.[442] "Similarly, the information we perceive in our consciousness is not created by conscious processes, nor is it reacted to by conscious processes. Consciousness is the middle-man, and it doesn't do as much work as you think. . . . We have long thought consciousness solved problems and had many moving parts, but it's much more basic and static."[443]

THE SHADOW BRAIN. Being aware of the unexpected ways a subconscious or unconscious human mind operates can give us a better chance of making rational judgements and decisions. *Photo by the author* (The painting is *"Study for Phidias in 'The Apotheosis of Homer'"* by Jean-Auguste-Dominique Ingres, San Diego Museum of Art)

While it may be disconcerting to stop and ponder the miniscule role "you" (your conscious self) have in your life, it certainly is not reason for panic or depression. Awareness here can be a good thing because it encourages second-guessing of important decisions and unusual perceptions. Learning that you have a silent but dominant partner living inside your head and weighing in on everything you think, feel, and experience can be powerful motivation to be diligent with critical thinking and skepticism. The role of the unconscious mind is yet another fundamental aspect of humanity that could be taught to everyone everywhere, including children, toward possibly uplifting the entire species. So many mistakes and tragedies are the result of that invisible and silent mind, the ninety-nine percent leading the way with little if any oversight. Most of the time the unconscious mind serves us well. There is no time for conscious reflections and deliberations when you touch a hot stove, for example. But it's a mistake to habitually and casually sign off in agreement with unconscious conclusions and commands during important moments. Deliberate conscious review is necessary when we think we see an alien spaceship in the sky, feel compelled to invest in a can't-miss opportunity, or think it makes perfect sense to go to war against some nation or people we fear. Being aware that we are never fully aware can go a long way toward making every life safer, more efficient, and more productive. We all make many mistakes over a lifetime. But some of the big ones might be avoided by simply pausing during the decision-making process to acknowledge the stealth role of the unconscious mind and review its directions.

When I was a child, I believed the world made sense. Although it was mostly mayhem and madness down at kid level where I was toiling away, I assumed that the grownups were doing much better up at their higher altitudes. As the years passed and my awareness grew, however, I realized how wrong I was. The world at large is something like a larger, dirtier and more dangerous version of my elementary-school playground. The lemmings, bullies, fanatics, and thieves I knew then never really went away. They just grew taller. While they no longer threaten knuckle sandwiches for lunch money, they now wage wars, poison the planet, and produce reality TV shows. Why is humankind so irrational, aggressive, and destructive?

Yes, we are uniquely intelligent and creative beings but we are exceptional for our madness as well. No other life-form on this planet makes up complex fictions and then proceeds to kill and die for them as we have throughout history. A key problem here is that the unconscious mind is too fast and too loose with the facts. In the name of speed and efficiency, it relies on partial, inaccurate, or inappropriate information to guide our decisions and actions. Fear and other emotions, rational or not, are routinely allowed to cast the deciding vote in what individuals think, say, and do. The result is the endless and global storm of prejudice, hate, and stupidity we see all around the world.

It can help the individual interested in being more rational to learn that many unconscious biases are constantly active in the human mind. These shortcuts aid or contaminate our thinking with little or no conscious participation and awareness. Regardless of our education level or intelligence, *no one* is ever completely immune from the constant push and pull of these biases. And while it would be difficult if not impossible for everyone to learn about every one of them, becoming familiar with even just a few of the most troublesome ones can clean up one's thinking to a significant degree. Confirmation bias, for example, is one of the most pervasive problems on Earth, and although I have noticed rising awareness of it in popular culture in recent years, most people still seem oblivious to it in their daily lives and continue to allow it to trip them up again and again.[444]

Confirmation bias is our natural tendency to notice, seize onto, and remember evidence that supports our beliefs while missing, ignoring, and forgetting evidence that opposes our beliefs. This is probably the most common reason why otherwise sensible people are able to confidently and sincerely hold nonsensical beliefs. It's not difficult to believe in a 6,000-year-old Earth or a flat Earth when one has a head full of facts and arguments in support of it but can't recall ever coming across any good opposing arguments. "The confirmation bias is one of the most insidious and pervasive bits of software in your head," psychology professor Hank Davis warns. "It is as much a part of being human as having two eyes, one nose, and two feet. To avoid evaluating the world through the confirmation bias, you have got to take conscious steps against it. Even then, there is no guarantee you'll succeed. If you allow your mental software to operate on its Pleistocene default settings, you will bring this bias into play."[445]

Confirmation bias is not only common but can be dangerous, too, sometime having life or death consequences. In healthcare, for example, the decisions doctors make when diagnosing and treating patients are susceptible to this bias and others. Jerome Groopman, a professor of medicine at Harvard Medical School, studied how the normal human biases of doctors might be harming patients. He found that *most* errors in healthcare are the result of unconscious factors such as confirmation bias rather than lack of expertise or technical knowledge.[446] Groopman cites a case of a doctor who had recently seen several patients with pneumonia and then misdiagnosed a woman. The doctor noted her symptoms that matched pneumonia but missed symptoms that contradicted it. It turned out the patient had aspirin poisoning. This suggests that *how* your doctor thinks may be as important as *what* your doctor knows. Groopman offers some practical advice to combat this danger. Try to stimulate more conscious oversight from a doctor by asking simple questions: "What else could this be?" or "What is the worst thing it could be?"[447]

The best way to fight confirmation bias in one's personal life is to consciously and deliberately make the effort to notice, pay attention to, listen, and sincerely consider opposing viewpoints and surprising evidence, or evidence that opposes currently held beliefs. Escape the filter bubble that may be presenting you with a biased, distorted view of reality. Pry open your mind and expose it to alternative views. Try to imagine ways in which your beliefs and conclusions might be wrong. Clearly, this is all easier said than done, of course, but definitely possible. Liberals can consume at least some news from conservative sources to broaden their view. Conservatives can do the same from liberal sources. Atheists and religious people might try harder to actually listen to one another. Perhaps the most important and helpful thing to remember is that humility is a prerequisite to good thinking. Never forget that every human being is loaded with stealthy biases that can lead them virtually anywhere, no matter how irrational. Resist and reject feelings of infallibility and absolute certainty. Think like a good scientist, i.e., consider everything, doubt everything, and follow the evidence.

THE GREEN FACTOR

Addressed in some detail in chapter 3 ("Who Are We?"), it is important to know that *the unconscious mind loves a green environment.* This makes sense because the primate brain evolved within wild settings over millions of years. Therefore, its natural environment is not a concrete box with a roof, a metal rectangle with four wheels, or an office cubicle. Although a brain and its host body may be physically safer in a suburban neighborhood as opposed to a wild prehistoric environment, an abundance of scientific research has made it clear that being in the presence of plant life and animals (excluding large predators, of course) can deliver *positive physiological and psychological benefits* to people. For example, a simple, brief walk among trees can help a person recover faster from mental fatigue. It can lower blood pressure, reduce stress, improve negative moods, elevate general cognitive abilities, and improve memory.[448]

Our relationship with the unconscious mind is complex. We need it, of course, because it does so much for us. We can't live without it. But anyone with a desire to be as sensible and safe as possible must respect its power while never completely trusting it. It's not easy, however. Part of the problem is that we even have an unconscious bias that keeps us from seeing our own unconscious biases. Appropriately named the Blind Spot Bias, it hinders us from easily recognizing our own irrationality, even as we see it in others around us.[449]

Consider the authority bias, another universal feature of the human mind, that every human has to contend with.[450] The unconscious mind would have us trust just about anyone with a fancy title or clothed in a well-pressed uniform. This is dangerous and can lead to problems, of course. Difficult as it may be for some to believe, there have been scores of incidents reported in the United States of someone phoning a business, pretending to be a police officer investigating a theft, and then instructing managers or other people to conduct strip searchers on female employees. One of these bizarre incidents took place at a McDonald's restaurant in Mount Washington, Kentucky. Based on nothing more than a voice on the phone claiming to be a police officer—a perceived authority figure—

an eighteen-year-old female employee was tormented and sexually abused in a back office for more than two hours. She followed the orders of her manager—an actual authority figure—to strip, show her vagina, perform jumping jacks while nude, and, incredibly, submit to nearly ten minutes of spanking while nude.[451] Such a ploy may seem too ridiculous to ever work. But it can because of a weakness that is a standard feature of the human unconscious mind. Virtually anyone can be surprisingly agreeable to following even the most outrageous orders from criminal or incompetent people so long as they perceive them to have authority.[452]

The Anchoring Bias is another deeply problematic unconscious bias.[453] It stems from the brain's admirable work ethic and noble desire to help. If unable to find relevant input to use toward coming up with an answer or conclusion, it won't give up. Instead it will drop anchor and use just about anything available, no matter how extraneous. For example, if I were to whisper, "billions", in your ear or discuss the number of stars in the Milky Way galaxy and then, later, ask you to estimate how many leaves are on a nearby tree, your answer would likely be a significantly higher number than if I had earlier whispered "seven" or discussed the number of people who were stranded on the television program *Gilligan's Island*. Because you don't have any idea how many leaves are on the tree, your subconscious mind would search for helpful input, of any kind, and recent exposure to a high or low number—however irrelevant— could be enough to shape the answer.

We might think more clearly and make better decisions if we remember that the unconscious mind has a tendency to make do with whatever information is available. This may seem absurd, impossible even, but it is typical of the unexpected activities of the unconscious mind. Don't let the Blind Spot Bias fool you into thinking that your brain is safe from the Anchoring Bias. There is plenty of solid research to show just how natural and ubiquitous it is.[454]

CUTTING THROUGH THE FOG

Thinking like a scientist can be a powerful defense against being misled by these biases. One of the reasons science works so well is that there is so much effort and so many procedures in place to identify, eliminate, or at least push back against unconscious bias. The randomized double-blind, placebo-controlled trial, for example, is called the "Gold Standard" because it does an excellent job most of the time in preventing unconscious biases from skewing experiments and leading to misinterpretations of results. The "double-blind" component, for example, typically means in that neither the researchers nor the test subjects know who gets the placebo and who gets the active drug or treatment in an experiment. If the researchers knew who got what they might unconsciously record biased or false results to steer the experiment toward an outcome their unconscious mind desires.

Authority, popularity, anchoring, and so much more might cloud many minds for a time but eventually, as we have seen again and again, the observations, experiments, and evidence tend to win out. Before making important decisions in life, the layperson can look to the core philosophy of science for help. Demand evidence, doubt, test, and never trust completely a human mind alone to determine reality.

Perhaps the most important thing to remember regarding the many problems presented by normal, healthy brains at work is that intelligence is not the solution. Far from providing a reliable defense, having high intelligence might make a person *more* irrational because he or she can use that marvelous mind to defend a crazy conclusion to no end with complex arguments and mountains of evidence collected by the confirmation bias. [455] Fortunately, the problem of unconscious biases and sloppy thinking in general has been getting more attention from researchers in recent years. There is even a call for the adoption of an "R.Q. test". Similar to the I.Q. test, the R.Q. test would attempt to measure a person's *rationality quotient*, how good she or he is at reasoning and thinking clearly.[456] The good news here is that a poor score would not be the end of it because we all can improve our thinking skills dramatically with more awareness, by applying critical thinking skills and strategies, and through specific training.[457] Science says, we can do better.

BELIEVING WHAT WE SEE AND SEEING WHAT WE BELIEVE

One of the most important scientific discoveries that everyone can benefit from knowing is that the human brain does not faithfully present the world to us as it really is. Were it widely understood, this one simple fact with profound implications might revolutionize and upgrade human culture. *We do not experience the world as it really is.* Only through conscious, deliberate, evidence-based thinking can we be somewhat sure of anything. Consider the countless mistakes, tragedies and deceits that came from this troublesome truth about the human brain. When you see, hear, or touch something your brain *creates* a perception that may be based on reality. It's not even meant to be reality but rather a useful representation of it. Based on some input, the brain puts together what it intuitively decides will be useful to you in the moment. Details that are judged on the fly to be irrelevant or distracting may be omitted from the finalized perception that goes live inside your head. If there are elements in a scene that your brain determines should be there but are not, it may fabricate them so that the final perception might be more useful to you. Further complicating things, your perceptions are shaped and colored by many unconscious factors. These can include your emotional state, physical comfort of discomfort, fears, beliefs, and desires. This system may work very well most of the time because it is a fast and efficient way to cope with normal day-to-day activities. But it is not a reliable way to discern truth from fiction when the stakes are high.

The reason millions of people around the world, generation after generation, can believe with total certainty that they have seen alien spaceships, ghosts, supernatural images in the sky, or other unlikely things is because they have not learned how the human brain perceives the world. They are ignorant of how their own brain operates and this leaves them highly vulnerable to drawing the wrong conclusions and forming irrational beliefs. Those who understand that the brain is not a camera or audio recorder more easily recognize the need for caution, humility, and skepticism in the face of extraordinary or odd experiences.

Everyone is likely to perceive weird things over a lifetime. What counts is how we process these moments and what we decide to take away from them. I once spent a week alone on a small deserted island in the Caribbean and during a windy night heard a voice even though no one was there. I once saw a streak of light in the sky moving so erratically that it didn't seem like it could be plane, meteorite, or rocket. But I don't believe in ghosts and UFOs today because I know enough about the human brain to know that it can and will fool me from time to time. It is possible that a ghost spoke to me that night on the island and maybe I really did witness a scout ship from another star system surveying Earth. But I doubt it. After being alone in silence on that island for days, it is more likely that my brain produced an audio hallucination or simply misinterpreted some natural sound for a human voice. The light in the sky could have been something real and normal that I observed accurately but didn't have sufficient knowledge to make sense of. Or, maybe my brain conjured up an exaggerated scene based on minimal visual input. Another possibility is that I do not remember either of those events accurately because memory is another significant challenge for every human.

DON'T FORGET THAT MEMORY IS UNRELIABLE

People who do not understand how human memory works are likely to have unwarranted confidence in their recollections and this can lead to many problems. Any person who has some basic awareness of what science has revealed about memory is probably going to be more careful about placing too much trust in their recall and avoid many mistakes over a lifetime. It is normal for us to feel very confident about a memory—we may clearly "see" a past event replaying in our minds—but there is a good chance that the memory is partially or totally inaccurate. Science informs us that confidence has little or nothing to do with accuracy when it comes to memories.

Human memory is complex and not completely understood. It doesn't happen in one specific place in the brain and memories are not sorted and filed in some orderly and logical sense the way a computer might. Our memory is *associative* and *constructive*. This means that one memory gets tied to another, or several, and the brain constantly works to find connections between stored memories and current activities or perceptions so that it can call up what it supposes is the most helpful memory for you in a given moment.

Memory is constructive because it pieces together a memory from input. Think of how an archaeologist builds a picture of the ways people of an ancient culture once lived. Bits of available data, called artifacts, are drawn together and a story of the past emerges. When it comes to memory the brain is like having your own onboard archaeologist. It uses artifacts from your past to *construct memories*. Further complicating things, however, is the fact that the brain is not committed to accuracy the way a professional archaeologist with high standards would be. It doesn't care about the memory being faithful to what really happened. It only wants to make that memory useful in the present toward making decisions about the future of the individual. Contrary to popular belief, nothing is being replayed in any that is like a video tape. Memories are reenactments with varying degrees of accuracy. Every brain is a theater in which loose interpretations of the past are performed for an audience of one. Never forget that the brain evolved for survival in the Pleistocene environment. It is of the prehistoric world, but finds itself bouncing around in an artificial, high-tech, crowded, cacophony of strange demands and distractions. We were hunters and gathers for most of our existence, farmers for a while, and now, in the most recent moment, we are expected to be something like court stenographers with photographic memories.

Most people do not fully appreciate just how unreliable human memory is.[458] They tend to assume that if a recollection feels accurate it must be accurate. To help people tune in to the reality of their brain's ways, I often share a very telling study that revealed just how far apart the public's common beliefs about memory are from the scientific view of it. First the public. This is what laypersons, people who do not study memory, think: Sixty-three percent believe that human memory "works like a video camera, accurately recording the events we see and hear so that we can review and inspect them later." Slightly more than thirty-seven percent believe that "the testimony of one confident witness should be enough evidence to convict a defendant of a crime." More than half of those surveyed by the researchers, 54.6 percent, believe that "hypnosis is useful in helping witnesses accurately recall details of crimes." And nearly half, 47.6 percent, of nonexperts believe that "once you have experienced an event and formed a memory of it, that memory does not change." Okay, so where do scientists who specialize in human memory come down on these same points? How many of the scientists surveyed agree that memory is like a video camera; one eyewitness testimony should be enough to convict; hypnosis can retrieve accurate memories and help solve crimes; and that once a memory is formed it stays unchanged in the mind? Zero percent. Not one expert agreed with any of these propositions. "In a sense, the results of this survey are disappointing," write Daniel Simons and Christopher Chabris, the psychologists who authored the study. "Many of the ideas we tested refer to scientific findings that have been established for decades. This discrepancy between science and popular beliefs confirms the danger of relying on intuition or common sense when evaluating claims about psychology and the mind."[459]

The reason memory is often problematic for us today is because we ask it to do too many things that it is not built to do well. One tries to recall an important conversation about salary with a boss a week ago or a year ago and expects the memory to be accurate. But there is no reason to trust it completely because the brain is not an internal scribe or voice recorder. Thanks to time, emotions, and current factors, any memory is likely to be edited, finessed, altered, compressed, or expanded. For this reason, with the benefit of scientific knowledge, we all can conduct our lives with a higher degree of accuracy and safety by consistently challenging and factchecking our memories on important matters. Know what the experts figured out through years of research and experimentation: The clarity of a memory in the mind and whatever confidence we may have in it, is just not enough to be sure it is accurate.

Another important aspect of memory is that it is meant to be *predictive*.[460] It works for us by serving up useful data from our past that might inform us as to what our next best move ought to be in the future. Most people probably think of memory as being all about the past, but this is just not the case. The brain is not even trying to store all past experiences for later retrieval because that is not in the brain's job description. Its duty is to decide what is worth storing, organize it, and then make it available when needed to enable us to *predict* what will happen if we do one thing as opposed to another. The brain's priority has never been to recall scenes from childhood or even the day before in great detail and with great accuracy. The priority has always been access to bits of information and glimpses of experiences from the past that can allow us to do a better of predicting the future.

Another disturbing though standard aspect of human memory is that it has been shown to be highly vulnerable to contamination or manipulation. It turns out that it is remarkably easy to get someone to confidently believe they remember something that happened, even though it never did. Nothing more than few well-timed words or even a single fake or misleading photograph can stimulate the brain to produce false memories that feel authentic. *Everyone* is vulnerable to this phenomenon. Elizabeth Loftus, a leading memory researcher and University of California–Irvine distinguished professor, warns that "manufactured memories are indistinguishable from factual memories."[461] Human memories are constructed and routinely *reconstructed*, Loftus explained in TED Talk. "Memory works a little bit like a Wikipedia page. You can go in there and change it— but so can other people. . . . We can't reliably distinguish true memories from false memories. We need independent corroboration."[462]

The brain is not your personal historian or journalist; it is your consultant. It draws on some general picture of the past to provide input toward making safe and productive in the present and in the future. The brain tells you *a story* about your past to help inform your next move. Accuracy, down to the word or digit, is not usually deemed necessary by the brain. Know who, where, and when you are. You belong to a species that is sloppy with details yet has constructed societies that are very detail oriented. We can compensate for this somewhat and help our brains help us by taking notes, keeping records organized and accessible, and—most important— always striving to stay humble and skeptical regarding important personal memories. I recommend thinking of a memory as a *starting point*, a sign post that can help us know something about that happened in the past but cannot be counted on as a trustworthy record of that past.

Chapter 10

HOW WILL EVERYTHING END?

W e just can't get enough of "The End". Religions have their doomsday prophecies. Hollywood sells their ninety-minutes of apocalyptic entertainment. And then there are the scientists, my personal favorites, who have a lot to say about how everything ends, too. The professionals who zero in on doom, horror, and spectacular cosmic finales from the scientific perspective observe, calculate, explore, experiment, and imagine so that we all can know some of the possible futures ahead. The general conclusion of this work: *We're all gonna die!* At some point in the near or long-term future, humankind will be severely tested if not summarily wiped out forever. But what will the threat be? When might it happen? Fortunately, science has a lot to say about this.

There is an excellent chance that human extinction won't occur any time soon. The average run for a mammal species is about a million years, which would seem to give us some breathing room.[463] But we are no typical mammal species. Humans are, for example, the first lifeform on this planet, maybe in the universe, to invent and build the means to bring about its own extinction—nuclear weapons. We also are unique for how fast and thoroughly we are altering or razing the natural global systems we depend on for food, water, and air.

Sooner or later, one way or another, *Homo sapiens* will go away. The current version of humankind may depart by smooth evolution or fiery destruction. It is only a question of when. I should note that nothing here is meant to terrify or panic. While reading this chapter keep in mind that death by age, injury, or disease is far more likely to take your life than some catastrophic event that extinguishes all of humankind. This does not mean, however, that we shouldn't give serious thought to evidence-based extinction events that could happen.

There are, of course, the practical reasons to consider. We need to protect and defend ourselves if possible. But there is also the scientific awareness factor. Knowing ourselves, living lives as awake as we can be and with as much meaning as possible, means learning about our origins, our surroundings, our present condition, *and all the terrible things that might befall us, too.* Studying threats to human survival in the future can even provide us all with some positive inspiration right now. Phil Torres is a scholar who focuses on existential risks, all realistic threats to human survival. He is the founder of The Project for Future Human Flourishing and the author of *Morality, Foresight, and Human Flourishing.*[464] His work incorporates both the practical and the profound aspects of the end. "From a philosophical perspective," Torres said, "that perennial—and profound—question of who we are cannot be answered without some sense of where we're going—what the future holds, the promises and pitfalls, the existential hazards and utopian rewards. But it may also be the case that what we do today, in this century, impacts the long-term future of our species in the universe. For example, we stand on the brink of colonizing space, the last great frontier, yet we also face a growing number of unprecedented risks to our survival. If humanity manages to overcome these risks and venture beyond this planetary spaceship called Earth, our descendants—and perhaps some people alive today, depending on when life extension technologies arrive—could realize literally astronomical amounts of value, filling our dark universe with life, consciousness, and hope."[465]

The end of humankind, Earth itself, and even the universe are not only fascinating topics to think about, but doing so could be key to saving our species one day. For example, thanks to science, we know that asteroids big enough to cause mass extinctions have hit Earth in the past and more will threaten our planet in the future. Most experts think a collision course is inevitable, given enough time, so we had better detect it and be ready to do something when that day comes. Smart people are working on this right now.[466] Hopefully, when the day comes, people will be ready to nudge the asteroid off course or do whatever defensive measures are deemed best. Whether it is giant space rocks or some other doomsday threat, being aware of our vulnerability and seeking knowledge about our own extinction, from the scientific perspective, can give us the motivation to do the necessary work so that humankind might avoid a premature extinction.

Donald Lowe, professor of geological sciences at Stanford University, believes investigating mass extinctions of the distant past are crucial to intelligently anticipating what might be lurking around the corner for us. "Mass extinctions have been a characteristic of the evolution of life, probably since the beginning," Lowe told me. "About 252 million-years-ago there was a mass extinction when perhaps 90-95% of all marine animals and a large proportion of land animals became extinct. Another mass extinction occurred 65 million years ago when the dinosaurs became extinct. Can it happen again? Could it happen tomorrow? How would we know unless we studied these ancient events to determine what caused the mass extinctions of the past? Without these past events and the efforts to understand them, it would all be speculation. In the case of the 252 [million-years-ago] extinction, it was probably volcanic magma reacting with coal deposits in [the region that is present-day] Siberia to release vast quantities of CO_2 and acidify the oceans, diving global anoxia. In the case of the 65 [million-years-ago] extinction, it was a large meteorite impact that had severe environmental effects. These are just few ways in which knowledge of the Earth, its make-up, the processes that affect its surface and interior, and its history affect humans and the human condition."[467]

Looking back helps us to see what is coming.

THE END OF HUMANS

The extinction of *Homo sapiens* will happen. This simple fact might might be jarring to some but it's true. Humans in their current form will not be around forever as one of two things will happen:

(1). *Homo sapiens* will become extinct as most species before us have. It will happen to us most likely by one or some combination of natural or human-generated scenarios. Such an event could happen in the distant future or tomorrow.

(2). *Homo sapiens* will evolve into something different enough to be designated a new species. This is fascinating because it could happen by natural means over a very long period of time, or it could occur quickly should we take a more direct and deliberate role in our own evolution. The rapidly growing capabilities and potential for genetic and artificial enhancements present the very real possibility that we may soon evolve ourselves into a new species. Either way, *Homo sapiens* is not a permanent feature of the universe. Charles Darwin recognized this more than 150 years ago, writing in *Origin of Species*: "Judging from the past, we may safely infer that not one living species will transmit its unaltered likeness to a distant futurity."[468]

There are many evidence-based threats to human existence, each with a different degree of risk. Global warming, for example, is a significant concern, at least for the scientifically literate, but I do not see it as a high extinction threat for humans. It may bring about the extinction of many other species, cause tremendous economic damage, and significant numbers of people could suffer and die, of course, but there is no credible global warming scenario I know of capable of killing off *Homo sapiens* within the next couple of centuries or so. It is important to keep in mind when thinking about extinction threats that it would take relatively few people to make a viable reproductive population. Cameron M. Smith, a Portland State University anthropologist, published a study in 2014 that looked at the minimal number of people needed to maintain a genetically healthy population for a theoretical long interstellar journey to an exoplanet. Smith concluded that it could be done with between 14,000 and 44,000 people. A "safe" population that accounts somewhat for illness and injury, he reasoned, would need to be about 40,000 people strong and include 23,400 reproductive males and females. "This number would maintain good health over five generations," Smith wrote, "despite (a) increased inbreeding resulting from a relatively small human population, (b) depressed genetic diversity due to the founder effect, (c) demographic change through time and (d) expectation of at least one severe population catastrophe over the 5-generation voyage."[469]

SUPERVOLCANOES

Supervolcanoes, though infrequent, pose a realistic threat to humankind. These subterranean monsters are *thousands of times* more powerful and destructive than regular volcanoes and are believed to be capable of causing mass extinctions. One could possibly terminate humankind at any moment by exploding and spewing so much smoke, ash, and dust into the atmosphere that the planet would be plunged into an instant ice age. Agriculture would fail and wild ecosystems would collapse quickly so there would be no resources to feed more than seven billion people. But while this would be terrible times for humankind, enough humans likely would survive it to keep the species in the game. There is a precedent. *Homo sapiens* survived at least one supervolcano already. It happened about 73,000 years ago when Toba erupted on the Island of Sumatra, Indonesia. As a result, the human population faced a "severe population bottleneck" and may have been down to only 3,000 to 10,000 individuals on the planet.[470]

One supervolcano lurks beneath Yellowstone National Park in northwest Wyoming. If it were to blow, its massive magma chamber could eject as much as 250 cubic miles of molten rock and ash into the atmosphere. This would cause a dark and cold global winter that would be devastating to most life-form. Yellowstone last erupted about 631,000 years ago.[471] It was widely believed that a supervolcano exploded somewhere on Earth about once every 100,000 years or so. However, a team of scientists at the University of Bristol published research in 2018 indicating that the frequency of eruptions is much higher, with one occurring perhaps every seventeen thousand years.[472]

ASTEROIDS

An impact from an asteroid large enough could easily wipe out *Homo sapiens* and most other Earth life as well. This is an interesting risk because it is extremely rare yet would be so catastrophic if it were to happen that it is not rational to ignore it. Earth has suffered significant impacts in the past and will be threatened with more in the future. Every planet and every moon in our solar system ends up in the crosshairs many times. Given enough time, it is inevitable.[473]

How big would the asteroid need to be to cause human extinction? The one that hit Earth 65 million years ago and killed most dinosaur species was about six miles or ten kilometers in diameter. It was big but one that size would not necessarily be enough to wipe out all of humanity today. To do that, to incinerate or obliterate everyone and leave no doubts, would take an asteroid that was about 60 miles wide, at least.[474] That may seem impossibly large but the solar system certainly is capable of serving up such a planet killer. Geologists have uncovered evidence of an impact occurring 3.26 billion years ago from an asteroid that was about *thirty miles wide* (50 kilometers).[475] The largest asteroid currently being tracked by NASA has a diameter of more than 300 miles.[476] There are even larger dwarf planets that could collide with planets, Earth included. Fortunately, no asteroids or dwarf planets currently are known to be on a course that threatens Earth.

FOSSILIZED WARNING FROM THE PAST. The iconic triceratops, like many other dinosaur species, was doomed to extinction when an asteroid six-miles (10k) wide struck Earth about 65 million years ago. *Photo by the author*

The danger posed by an asteroid, meteoroid, or comet is not contingent on size alone. The kinetic energy (energy of motion) can be so great that something most people probably would think of as a very small object can still deliver an enormous blast. Asteroids typically move at speeds of tens of thousands of miles per hour. According to NASA, twelve miles per second is about average speed.[477] Therefore, an object that is ten-meters wide (33 feet) can still pack the approximate power of five of the nuclear bombs dropped on Hiroshima during World War II.[478]

NASA brought together a panel of experts to look into the threat and their conclusion was that it demands attention and advance planning: "Although the annual probability of the Earth being struck by a large asteroid or comet is extremely small," the researchers declared in their final report, "the consequences of such a collision are so catastrophic that it is prudent to assess the nature of the threat and to prepare to deal with it."[479]

One of the key findings of the NASA study group was that an object around two kilometers in diameter would be enough to alter global climate and cause widespread crop failures and famine. It would be as if a hundred of the most powerful nuclear weapons every built all exploded at once. This would, they suggest, not be enough to necessarily cause human extinction. It would, however, likely mean the death of more than a billion people and the collapse of civilization as we know it. The NASA team calculated that such impacts seem to occur on Earth from one to several times per every one million years.[480]

Assuming we detect an incoming asteroid, what can be done about it? Best-case scenario would give scientists and governments several years of warning. Most experts favor one of two actions. One is the "kinetic impactor." This means launching something at it (not a bomb) and slamming into it. If it's far enough out, this will slow it slightly and alter its course just enough to make it miss Earth.[481] Remember that the Earth is also always moving at a very high speed. Another population option is a "gravity tractor." This involves stationing something close to the asteroid that has enough mass—and therefore gravitational pull—to slightly alter the asteroid's course. Simply firing a nuclear missile at an incoming asteroid is generally considered to be a bad idea. As hard as it is to believe, even the most powerful nuclear warheads available today could not stop the mass of a large asteroid travelling at twelve miles per second.[482]

ARTIFICIAL INTELLIGENCE

The pace of computer progress has been astounding in recent decades. We are going somewhere fast. But while the potential benefits of superintelligence are great, there are dangers as well. While the scenario presented by the Terminator films—armed robots loose in the streets zapping fleeing humans—may be unlikely, many bad things could happen should one computer or network of computer attain god-like powers over us.

People considered by most to be worth listening to on high-tech matters have serious worries. Bill Gates, Steve Wozniak, Nick Bostrom, and Elon Musk, for example, have expressed concerns.[483] "I have exposure to the very cutting-edge AI, and I think people should be really concerned about it," said Musk in 2017. "I keep sounding the alarm bell, but until people see robots going down the street killing people, they don't know how to react, because it seems so ethereal." Musk is pushing for immediate government regulations on AI research. "AI is a rare case where we need to be proactive about regulation instead of reactive. Because I think by the time we are reactive in AI regulation, it's too late. AI is a fundamental risk to the existence of civilization." [484]

Stephen Hawking, the prominent theoretical physicist, seems resigned to a defeat on this issue. "Once humans develop artificial intelligence," Hawking told the BBC. "it will take off on its own and redesign itself at an ever-increasing rate. Humans, who are limited by slow biological evolution, couldn't compete and would be superseded."[485] University of Oxford professor Nick Bostrom is director of the Strategic Artificial Intelligence Research Centre. He too thinks we are heading for trouble: "Before the prospect of an intelligence explosion, we humans are like small children playing with a bomb. . . . Superintelligence is a challenge for which we are not ready now and will not be ready for a long time. We have little idea when the detonation will occur, though if we hold the device to our ear we can hear a faint ticking sound."[486]

One of the most concerning challenges concerning AI is that we aren't very good at predicting the future. A superintelligent machine may do things its creators or owners never anticipated because it thinks in different ways and has access to more data. Anca Dragan, an assistant professor at UC Berkeley, believes that good intentions mean little when it comes to AI. "Even if we plan to use AI for good," she warns, "things can go wrong, precisely because we are bad at specifying objectives and constraints for AI agents. Their solutions are often not what we had in mind."[487]

Long-term, I think the machines win out. However, I total subjugation or annihilation by machines is our likely fate. Nor, however, do I see the computerized utopia some envision happening anytime soon. I predict a murky mix of both good and bad outcomes. We will continue to see benefits to healthcare and transportation in the near future, for example. But increasingly powerful computers will also enable bad people, companies, and governments to do very bad things to people. As algorithms become more complex and capable, the ambitions of those who own them are sure to rise too. We are likely to see, for example, AI used to manipulate the minds of mass numbers of people in ways previously confined to the pages of science-fiction novels. We already see this happening with social media.[488] In the end, the machines may get us, but for now I find more reason to fear the government, corporation, or terrorist group in possession of a computer with AI than a computer alone. There may also be reason to worry about dumb machines, too. Nanotechnology is a potential technological threat that doesn't get as much attention these days. It seems incredibly far-fetched, but self-replicating, matter-consuming nanobots could theoretically escape human control, or be used maliciously by humans, and possibly destroy humankind.

Finally, the danger of an AI takeover could be eliminated by the possible—some would say inevitable—merging of human and machine. This already is well underway, of course, with various medical implants and prosthetics enhancing lives around the world. Facebook has invested some of its wealth into developing a brain-computer interface that would enable people to type with thoughts at a rate of 100 words per minute.[489] Musk, the founder of Tesla and SpaceX, is presently attempting to construct a "direct cortical interface" that would be implanted in the brain so that people can "upload and download thoughts."[490] It is difficult to imagine how the machines will rise up against us if we are the machines.

<BOX>

CONFRONTING THE FATE OF THE UNIVERSE

Ethan Siegel is a theoretical astrophysicist and science writer with a supernova personality and flair for making complex science fun and appealing to anyone with a pulse. He is the author of *Beyond the Galaxy: How Humanity Looked Beyond Our Milky Way and Discovered the Entire Universe* and *Treknology: The Science of Star Trek from Tricorders to Warp Drive.* I asked Siegel about the beginning and end of everything.[491]

Image: "Ethan Siegel" Ethan Siegel *Photo by Jamie Cummings/John Morgan*

WHY DOES LOOKING UP THE STARS MATTER AND WHERE DOES YOUR DESIRE TO KNOW THE UNIVERSE COME FROM?

I grew up in New York City and didn't have access to dark-night skies. But some of the best memoires I have as a child were camping in the summer, maybe by the Catskill Mountains, maybe by Lake George. At night, things were the most spectacular to me because you would see the night sky open up. In New York [City] the sky never gets completely black and at the most you could see a dozen or two dozen stars. Out in the country you lay down in the grass and look up, it just looks like there is a whole universe out there. And when you start to think about it and wonder about it, you realize, wow, each one of those points of light is a sun just like our own, and maybe they have planets, and maybe it goes on for infinity. All these questions you wonder about as a child, and then find out that science has actually made huge strides toward actually finding the answers to. When I realized this, it blew my mind and I just yearned to know as much as I could about what's out there.

You communicate science to the masses very well. But why do you do it? What is your goal? Do think scientific awareness enhances people's lives?

When most people talk about science education or science communication, they talk about what people "should know" about these things or how everyone should be able to "solve these problems" or "answer these questions". I don't go in for any of that. I think the best thing you can do for people is to give them an awareness of what science *is*, how we obtain scientific knowledge, and an appreciation for what we know, how we know it, and how it makes our lives better.

What is science?

Science is not just a collection of facts or various properties to memorize. Science is both the full suite of knowledge that we've gathered about a topic and it's also a process. It has unique value that everyone should value, that everyone should share in, and reap the fruits of. Science gives us a uniquely powerful way to find things out and to know them with a level of certainty that no other field of inquiry will reproduce. It's a way of gaining information about the world. It's a way of learning what's out there in the universe by asking the universe questions about itself. There is nothing else that accomplishes all this the way that science.

What is something that would surprise most people about the Big Bang theory?

It wasn't the beginning of everything. There must have been something that happened before you get to that singular point that set up all the initial conditions of the universe. Somehow, something must have happened.

Some people see current gaps in knowledge as a flaw or weakness. How would you like people to think of unanswered questions in science?

Unanswered are part of the nature of science. Unanswered questions are what happen when you look to the universe, ask it questions, and collect data. That answer, whatever it is, will inevitably raise new questions. There is never a final answer. There will always be more questions. . . . Science isn't about finding the perfect answer to all your questions. Science is about investigating the natural world to the best of your abilities. And the better you get at it, the more you do it, the deeper your questions can be. Science is something that will go on and on and on as long as there are things to learn about the universe, and that should be forever.

What does science say about the end of everything?

The universe hasn't been this way forever and it won't be this way it is now forever. We look out and see stars, and galaxies, and a whole slew of things that are chances at life, chances at intelligent civilizations, and chances to be weirder and more cosmically awesome than we ever imagined. But it won't be this way forever. The fuel that powers the stars, hydrogen, gets burned up with each new generation. The stars, as long lived as they are, will not live forever. Our Sun is going to die. It looks like the majority of stars that will ever exist in the universe have already formed. Because the universe is expanding, and that expansion rate is speeding up, everything that is outside of our Local Group (system of neighboring galaxies), all the clusters and superclusters out there will never become gravitationally bound to us. Eventually they will recede away from our view entirely. And even though our Local Group will merge into a giant galaxy, over a long enough timescale all of those stars will burnout. All of that material for creating new stars will get burned away. And the stars themselves will be gravitationally ejected from whatever our galaxy becomes. So, when we look to the distant future, there will be a **last star** that forms in our galaxy. There will be a central black hole that devours everything that isn't ejected and then it will decay. It will take an extremely long time, but all galaxies will suffer the same fate as our own.

What does this all mean for us? I think it means, on one hand, you can be reassured that no matter how poorly you mess up anything in this life, the universe is going to continue to unfold exactly as it was going to, irrespective of anything you have done. But second off, it also tells you that the universe has a long, rich life ahead of it before it become crashing down. And for that it's worth enjoying every minute of it, and this makes it all the more important to be good to one another for all the time that we are here.

NUCLEAR WAR

Throughout the Cold War (1947-1991) the threat of nuclear war possibly causing the end of humankind was something most people thought about—and for good reason. Nuclear war could have happened and almost did many times.[492] The stockpiles of these powerful, city-destroying weapons were absurdly huge. The Soviets amassed 40,159 nuclear weapons at their highpoint in 1986.[493] The United States nuclear arsenal peaked in 1967 at 31,255 weapons.[494] Since the collapse of the Soviet Union, however, nuclear weapons have largely faded away as focus of serious concern for most. There are the occasional spikes of concern: *What if a terrorist organization gets a nuke? Will North Korea launch a nuclear missile on Guam, Hawaii, or the continental United States?* But very few people these days seems appropriately concerned about the global threat to *Homo sapiens* posed by the *thousands of nuclear weapons* still waiting to bring down civilization should one or a few people with the power give the command to launch.

WHO HAS NUKES?

**Nine countries currently possess
approximately 15,000 nuclear weapons.**

Russia: 6,800 warheads
United States: 6,600
France: 300
China: 270
United Kingdom: 215
Pakistan: 120-130
India: 110-120
Israel: 80
North Korea: 10-20

Source: Federation of American Scientists 2017[495]

A remarkable study was published in 2014 that showed the devastating effects a relatively small nuclear war between two nations could have not just on their societies but on the entire world. The researchers projected what would happen if India and Pakistan each detonated fifty 15 kiloton weapons.[496] The study concluded that the explosions and resulting fires would send about 5.5 million tons of black carbon into the stratosphere that would spread around the planet. This would cause a rapid drop in surface temperatures and an intense heating of the stratosphere. The scientists predict global ozone reductions of twenty to fifty percent over population areas—unprecedented in human history—that would cause great harm to the health of people, and agriculture, as well as land and water ecosystems.

"These results illustrate some of the severe negative consequences of the use of only 100 of the smallest nuclear weapons in modern megacities," write the researchers. "Yet the United States, Russia, the United Kingdom, China, and France each have stockpiles of much larger nuclear weapons that dwarf the 100 examined here. Knowing the perils to human society and other forms of life on Earth of even small numbers of nuclear weapons, societies can better understand the urgent need to eliminate this danger worldwide."[497]

0.025 SEC.
N

100 METERS

DESTROYER OF WORLDS? The eerie fireball from history's first detonation of a nuclear weapon swells in the night sky of the Jornada del Muerto desert, New Mexico at 5:29 am, July 16, 1945. Today's nuclear weapons are thousands of times more powerful. *Photo U.S. Department of Defense*

Every society on Earth would feel the effects of a limited or full-scale nuclear war, no matter how far away from the combatant nations it may be. Given the scenario of a large exchange of weapons, the nuclear winter effect, collapse of trade and food sources, and pervasive radiation could possibly result in human extinction. I am doubtful, however, that anything short of a major exchange between the two biggest nuclear powers by far, the United States and Russia, would be enough to do that. Don't forget, *Homo sapiens* is an extremely intelligent, creative, adaptable and therefore durable species. There is even some science to support my strained optimism here. An interesting study that surveyed the worst possible catastrophes and likelihood of human extinction concluded that no one of them alone would likely be enough to end us for good. "The human race is unlikely to become extinct without a combination of difficult, severe, and catastrophic events," said lead researcher Tobin Lopes.[498] "[We] were very surprised about how difficult it was to come up with plausible scenarios in which the entire human race would become extinct." Camron Smith's study cited previously in this chapter put the minimal range of a viable human population from fourteen to forty-four thousand people. But that was based on a well-planned long-term space mission to settle an exoplanet and considered the need to be able to withstand a disaster along the way or after landing. In the case of an ultra-lethal viral plague, supervolcano explosion, asteroid impact, or nuclear war, humankind probably could scrape by with much less given no other choice. I suggest that a human population of less than five thousand, perhaps as few as two or three thousand people, could be sufficient to keep humankind going. There likely would be serious genetic issues from inbreeding, of course, but that's certainly preferable to extinction. Therefore, it is not difficult to imagine a few tiny pockets of survivors here and there, from a population of more than seven billion, making it through the catastrophe with intelligence, creativity, toughness, technology, and luck. Some scholars point to promising trends of peace and nonviolence, but how good can we justifiably feel about the future of *Homo sapiens* over the next hundred or two hundred years? [499] We are, after all, the species that continues to poison and eradicate ecosystems that sustain it; increase social/economic inequality in a fast-growing population; and continually invent more effective and convenient ways to kill human beings.[500] Next up: autonomous killing machines that walk, roll, swim, and/or fly.[501] What makes this behavior so maddening and frustrating is that we have the necessary information to guide us. Science has illuminated the safer path forward, but we have not taken it with appropriate haste. Humankind is shackled by its past. We are slowed, if not effectively immobilized, by old, outdated institutions; conflicting national, racial, and religious loyalties; and the mistaken perception that we are apart from nature. Molecular biologist Richard Lathe believes that the greatest threat facing humankind is humankind. "We are facing a future where the number of people on the Earth exceeds the capacity of the world to feed them," he said. "Ergo, famine, wars, and disease the like we have never seen before." The hope is that we will mature and decide to conduct ourselves in a sensible and sustainable manner before we run ourselves into a wall. Read the following quote from an essay by Stephen Hawking that was published in *The Guardian* in 2016. See if you can sense his hope for humanity struggling against his fears of the worst. It's an internal duel, one I am familiar with. "For me, the really concerning aspect of this is that now, more than at any time in our history, our species needs to work together." Hawking continues: "We face awesome environmental challenges: climate change, food production, overpopulation, the decimation of other species, epidemic disease, acidification of the oceans. Together, they are a reminder that we are at the most dangerous moment in the development of humanity. We now have the technology to destroy the planet on which we live, but have not yet developed the ability to escape it. Perhaps in a few hundred years, we will have established human colonies amid the stars, but right now we only have one planet, and we need to work together to protect it."[502]

In 2003 I interviewed Joseph Rotblat, winner of the Nobel Peace Prize for his decades of work to make the world safer from nuclear war. Rotblat, who died in 2005, had been one of the lead scientists in the Manhattan Project, the secret U.S. World War II effort to invent and build the first nuclear bomb. Rotblat told me that he only participated in the effort because he felt it was necessary for the United States to beat Hitler to the bomb. When he learned in 1944, however, that the Germans had no chance of developing nuclear weapons he resigned from the Manhattan Project. A year later, when two atomic bombs were dropped on Japanese cities, Rotblat said he was horrified and immediately dedicated his life to fighting against the very weapons he helped to invent. "Since the end of the Cold War so many people believe that nuclear weapons are no longer a threat," he said. "We have to keep reminding them. You must also remember that while we are advocating the elimination of weapons of mass destruction because they are a source of immediate danger, you must also realize that the removal of these weapons will not bring absolute security to the world. We cannot uninvent nuclear weapons. Even after we get rid of them, nations that have serious conflicts can rebuild these nuclear arsenals and we would be back in a Cold War situation. We must attack the institution of war itself. We must abolish war as a means of solving disputes. . . . There is no alternative. If we should not solve these problems then the end of civilization, the end of humankind, is much more likely to happen."[503]

I also interviewed Edward Teller, "Father of the Hydrogen Bomb," in 2002, one year before his death. Hydrogen bombs, or H-bombs, were developed in the 1950s. They work by nuclear fusion and are much more powerful than the fission bombs used to attack Japan. Speaking with Teller at length fascinated me because he was a rare person who had made significant contributions to the means of our possible extinction. Teller told me that he believed fear and the lack of respect for knowledge to be the greatest dangers to the world. When I asked him for his thoughts on our chances of long-term survival, he replied in his dramatic monotone way: "The fate of mankind is most uncertain."[504] Yes, it certainly is.

The most tragic outcome for *Homo sapiens* would be for our perpetual avalanche of short-sighted behavior and irrational thinking to finally catch up to us and exact the ultimate price. This could happen. Self-destruction or technological suicide is not another wild apocalyptic fantasy. The most intelligent lifeform to ever live on this planet really could kill itself.

GAMMA RAY BURST

Something happens about once per day in the universe that is so shocking and powerful that it hardly seems possible. When some stars die they discharge a burst of energy that can be the equivalent of *all the energy released by a thousand Suns over their entire lifetimes*. A star that does this can briefly outshine the universe and be seen thirteen billion light years away.[505] Called a gamma ray burst, this phenomenon is incredibly destructive. Relatively narrow beams of energy shoot out and obliterate whatever is unfortunate enough to be in its path. One gamma ray burst spotted in 2008 was bright enough to be visible to the *naked eye* here on Earth—even though it was nearly eight billion light-years away.[506]

Astronomer Phil Plait describes what a gamma ray burst occurring "only" a hundred light-years away from Earth would be like for us. Spoiler alert: Everything dies. The energy beam would travel toward as light speed and would be about *fifty trillion miles wide* so not only would the entire Earth be hit; it would blast the entire solar system as well. The average gamma ray burst duration is about ten seconds, so Earth could be under fire for a few minutes or less than a second.[507] Even though a brief burst would only hit part of Earth's surface because of its rotation, Plait concludes that no place would be perfectly safe and little life would be spared. "It would be like blowing up *a one-megaton nuclear bomb over every square mile of the planet facing the GRB*," Plait estimates. "It's (probably) not enough to boil the oceans or strip away the Earth's atmosphere, but the devastation would be beyond comprehension. Mind you, this is all from an object that is *600 trillion miles away*."[508] There is also the radiation to content with. Once the ozone layer is destroyed, the Earth and all exposed life would be bathed in lethal radiation.

Astronomer Chris Impey paints the following unpleasant picture of fiery agony should a gamma ray burst hit Earth: "Forests would burn, lakes and rivers would boil off, and the side of the Earth that faced the blast would be sterilized. The shock wave from the impact would send a mile-high wall of flames around the planet, perhaps sparing some ocean organisms on the far side." I love the optimistic offering at the end of that. Maybe, just maybe, some microbes would make it through even that. The good news is that we do not currently know of any stars that seem to threaten us with a gamma ray burst. The bad news, however, is that we do not know enough about all the stars in our galaxy and those beyond to declare ourselves safe. But don't lose any sleep fearing a gamma ray burst. If one were to go off with Earth in its cross-hairs, there would be no warning, nothing we could do about it anyway, and death would come quickly for us. Impey offers this contemplative take on human extinction via gamma ray burst: "For endings, it's hard to beat quick, painless, and high in drama. If humans are taken out by a gamma ray burst, it would be nature's biggest good-bye, visible across the universe."[509]

Unsettling as it may be, it is informative, enlightening, and necessary to think about the destructive power of a gamma ray burst, an event that happens on average once per day somewhere. This is the environment that we live in. It may be impossible for us to ever truly comprehend the universe, given the immense distances, size of objects, and destructive power it contains, but we must try if we hope to know our home and all possible futures.

THE END OF EARTH

At 4.54 billion years old, Earth has been around for a long time. But it won't last forever. Possible big problems for life on Earth one day include the previously mentioned giant asteroid hitting and sterilizing the entire planet. Or a nearby star might die in a spectacular supernova that dooms Earth with a shower of lethal radiation. There is no doubt that the Sun will eventually destroy Earth. No star is immortal. They are born, they live, then they die. The Sun doesn't have enough mass to go supernova but it will not go peacefully from Earth's perspective. In a few billion years the Sun will burn through its hydrogen fuel and the core will heat up, making it a significantly hotter and brighter star, so much so that Earth and all its residents will feel it in a big way. The polar caps will melt, the oceans will evaporate, and much of Earth's water will vent into space because the atmosphere will be too warm to contain it.[510]

As the Sun continues its long, hot death march it will lose surface gravity. This means the planets' orbits will expand outward because the Sun's hold on them is weaker. Scientists are not sure if Earth will be consumed by the still blazing hot Sun as it drifts further out or if it will escape that fate by staying just ahead of its reach.[511] By the way, all animal and plant life will be long gone. If *Homo sapiens* were to have intelligent descendants still around when the Sun begins its death throws, there is a chance they might be technologically advanced enough to migrate to a more suitable planet and avoid the worst of it. We can assume, however, that the first life on Earth will be the last life on Earth. The microbes, especially tough and ubiquitous bacteria, will hold on and probably be the last life standing.

THE END OF THE UNIVERSE

Not so long ago it was widely thought by scientists that the universe was infinite and generally static. It seemed that there was no end to it in space nor time, and that the universe would always be the way it is. Things have changed dramatically over the years, of course. Now we know that the universe is not only far from stagnant but rapidly expanding. We also have reason to suspect that it may come to a very definite conclusion one day in far, far future. So, if we do have intelligent descendants who manage to figure out a way to survive the death of the Sun in four or five billion years, then their descendants would later face the even greater challenge of trying to survive a dying universe.

Scientists are not in agreement about what may happen but there are intriguing possibilities that are much more than wild hunches. For example, the universe is not just expanding; it's doing so at an accelerating rate. And, not only are the galaxies, stars, and planets spreading out and away from each other, but space-time itself—the fabric of existence—is stretching apart, as well. This means that many billions of years in the future the universe may tear itself apart. Everything—every star, planet moon, rock, and lifeform—would be stretched apart and left to drift as a kind of thin mist of particles, maybe. Even atoms might not survive. Nothing will work, nothing will happen. Known as Heat Death or the Big Freeze, this seems like a rather sad end for the action-packed spectacular universe we know today.

Another possibility in the far future is that the universe might experience something like the Big Bang in reverse. If the expansion should slow and stop then gravity would have its opportunity to seize the moment and start the process of everything coming back together. Imagine that: all the contents of the universe contracting inward, rushing toward another tiny singularity point. This hypothesis is called The Big Crunch. Some cosmologists suggest that there might be many universes "out there" and every now and then two of them bump into each other, possibly creating new universes from the collisions. In 2015, *New Scientist* magazine reported on a strange glow detected out in the far fringes of the universe by the European Space Agency's Planck telescope. One lead scientists looking at the data thinks it might be evidence of two universes colliding or nudging by one another. It might even be energy from another universe leaking into ours.[512]

We should want to figure out as much about the fate of the universe as we can because this is our home and knowing its future will help us understand it now. We don't need to worry about the ultimate end because it's many billions of years away and we have more pressing problems. What I love about intellectually tussling with the greatest doomsday of all is that doing so might just represent the single most spectacular moment a human mind can enjoy. Even if the hypothesis we consider in a given moment is wrong, even if we don't fully grasp the concepts as intended by those who present them, we engage ourselves in something grand and meaningful. We use a human brain—this young, patchwork organ, that is best for helping primates find food and water, maintain group relationships, achieve sexual intercourse, imagine the next best tool—and to time travel to the end of time. Those who care to can ride this brain, with science as our guide, to the ends of the universe and beyond. With human brains we look forward ten or twenty billion years, *a trillion years*, and imagine every star dead, all planets devoid of life, and every atom dismantled and inert. With a brain we can see anything, and even nothing. Scientific knowledge adds excitement and richness to a life. Because I have learned basic information about the universe, the Earth, and life, I see things everyone can, but many do not. I look at a rock, a tree, or another person and see billions of years of travel and never-ending layers of exciting complexity. Where ever I go, I am surrounded by time capsules filled will wonderful and stunning stories. Science tells these stories. There are hundreds of billions of stars in our galaxy and hundreds of billions of galaxies in the universe. How can anyone ever be bored or lack inspiration? All we have to do is look up and contemplate the reality we know while continuing to explore. Try to find some joy in the good fortune of existence now. We live in good times, somewhere between the beginning and the end of everything. We know, or can know, enough detail to be excited and inspired by our time and place. We are surrounded by, inside of, and thoroughly infiltrated by mysterious and wonderful universes. There is the life that is all around us here on Earth, most of which awaits discovery and study.[513] Even the majority of the life-forms that live inside of us are unknown.[514] Venture close enough and matter, the stuff of existence, becomes a bizarre rabbit hole filled with surprises. And then there is the greater universe of stars and galaxies, of course, with its many stories still to tell. To tap into all this, we only have to appreciate the productive power of the scientific process, use our brains as best we can, and never stop exploring. The more we look, the more we will see. The more we learn, the more alive and awake we become.

BIBLIOGRAPHY

Ackerman, Diane. *The Human Age: The World Shaped by Us*. New York, NY: W.W. Norton, 2014.

Aczel, Amir D. *Entanglement*. New York: Plume ,2003.

Angier, Natalie. *The Canon: A Whirligig Tour of the Beautiful Basics of Science*. New York, NY: Houghton Mifflin, 2007.

Ben-Barak, Idan. *The Invisible Kingdom*. New York, NY: Basic Books, 2009.

Berger, Lee and Hawks, John. *Almost Human: The Astonishing Tale of Homo naledi and the Discovery That Changed Our Human Story*. Washington D.C.: National Geographic Partners, 2017.

Bostrom, Nick. *Superintelligence: Paths, Dangers, Strategies*. Oxford, UK: Oxford Press, 2014.

Brockman, John, ed. *What Have You Changed Your Mind About?* New York: Harper Perennial, 2009.

Brockman, John, ed. *What to Think about Machines That Think: Today's Leading Thinkers on the Age of Machine Intelligence*. New York: Harper Perennial, 2015.

Buonoman, Dean. *Your Brain Is a Time Machine: The Neuroscience and Physics of Time*. New York, NY: W.W. Norton, 2017.

Burton, Frances D. *Fire: The Spark that Ignited Human Evolution*. Albuquerque, NM: University of New Mexico Press, 2009.

Burton, Frances. *Fire: The Spark That Ignited Human Evolution*. Albuquerque: University of New Mexico, 2009.

Carroll, Sean. From *Eternity to Here: The Quest for the Ultimate Theory of Time*. New York, NY: Dutton, 2010.

Carroll, Sean. *The Big Picture: On the Origins of Life, Meaning, and the Universe Itself*. New York, NY: Dutton, 2016.

Cavalli-Sforza, Luigi Luca; Menozzi, Paolo; Piazza, Alberto. *The History and Geography of Human Genes, (Princeton, NJ: Princeton University Press, 1994.*

Chamovitz, Daniel. *What a Plant Knows: A Field Guide to the Senses*. New York, NY: Scientific American, 2012.

Cooper, Chris. *Everything You Need to Know About the Universe*. San Diego, CA: Thunder Bay Books, 2011.

Corfield, Richard. *Lives of the Planets: A Natural History of the Solar System*. New York, NY: Basic Books, 2007.

Coyne, Jerry. *Why Evolution is True*. New York, NY: Viking, 2009.

Davis, Hank. *Caveman Logic: The Persistence of Primitive Thinking in a Modern World*. Amherst, NY: Prometheus Books, 2009.

Davis, Hank. *Caveman Logic: The Persistence of Primitive Thinking in a Modern World*. Amherst, NY: Prometheus Books, 2009.

Dawkins, Richard. *The Selfish Gene*. Oxford: Oxford University Press, 1989.

———. *The Blind Watchmaker: Why the Evidence of Evolution Reveals a Universe without Design*. New York: W. W. Norton, 1996.

———. *Climbing Mount Improbable*. New York: W. W. Norton, 1997.

———. *The Ancestor's Tale: A Pilgrimage to the Dawn of Evolution*. Boston: Houghton Mifflin, 2004.

———. *The Greatest Show on Earth: The Evidence for Evolution*. New York: Free Press, 2009.

———. *The Magic of Reality: How We Know What's Really True*. New York: Free Press, 2011.

Dennett, Daniel C., *From Bacteria to Bach and Back: The Evolution of Minds*. New York, NY: W.W. Norton, 2017.

Everett, Daniel L. *How Language Began: The Story of Humanity's Greatest Invention*. New York, NY: Liveright, 2017.

Fairbanks, Daniel J. *Everyone is African: How Science Explodes the Myth of Race*. Amherst, NY: Prometheus Books, 2015.

Fuentes, Agustin. *The Creative Spark: How Imagination Made Humans Exceptional* (New York, NY: Dutton, 2017),

Fortey, Richard. *Life: A Natural History of the First Four Billion Years of Life on Earth*. New York, NY: Vintage Books, 1998.

Frans De Waal. *The Age of Empathy: Nature's Lessons for a Kinder Society*. New York, NY: Harmony Books, 2009.

Frazier, Kendrick. *Science under Siege: Defending Science, Exposing Pseudoscience*. Amherst, NY: Prometheus Books, 2009.

Freedman, Carl. *Conversations with Isaac Asimov*. Jackson: University Press of Mississippi, 2005.

Garreau, Joel. *Radical Evolution: The Promise and Peril of Enhancing Our Minds, Our Bodies—and What It Means to Be Human*. New York: Doubleday, 2005.

Fuentes, Agustin. *The Creative Spark: How Imagination Made Humans Exceptional*. New York, NY: Dutton, 2017.

———. *Race, Monogamy, and Other Lies They Told You: Busting Myths about Human Nature*. Oakland, CA: University of California Press, 2012.

Gott, J. Richard; Strauss, Michael A.; Tyson, Neil DeGrasse. *Welcome to the Universe: An Astrophysical Tour*. Princeton, NJ: Princeton University Press, 2016.

Gould, Stephen Jay. *The Flamingo's Smile: Reflections in Natural History*. New York: W. W. Norton, 1985.

———. *Ever Since Darwin: Reflections in Natural History*. New York: W. W.

Norton, 1992.

⸻. *Hen's Teeth and Horse's Toes: Further Reflections in Natural History*. New York: W. W. Norton, 1983.

⸻. *Wonderful Life: The Burgess Shale and the Nature of History*. New York: W. W. Norton, 2007.

Greene, Brian. *The Fabric of the Cosmos: Space, Time, and the Texture of Reality*. New York: First Vintage Books Edition, 2005.

Greene, Brian. *The Hidden Reality*. New York, NY: First Vintage Books, 2011.

Grossman, David. *On Killing: The Psychological Cost of Learning to Kill in War and Society*. New York, NY: Back Bay Books, 2009.

Guterl, Fred. *The Fate of the Species: Why the Human Race May Cause Its Own Extinction and How We can Stop It*. New York, NY: Bloomsbury, 2012.

Hanlon, Michael. Eternity: Our Next One Billion Years. London: Macmillan, 2009

Hanson, Thor. *The Triumph of Seeds: How Grains, Nuts, Kernels, Pulses, and Pips Conquered the Plant Kingdom and Shaped Human History*. New York, NY: Basic Books, 2015.

Harari, Yuval Noah. *Homo Deus: A Brief History of Tomorrow*. New York: Harper, 2017.

Harari, Yuval Noah. *Sapiens: A Brief History of Humankind*. New York: Harper, 2015.

Harrison, Guy P. *50 Popular Beliefs That People Think Are True*. Amherst, NY: Prometheus Books, 2012.

Harrison, Guy P. *Good Thinking: What You Need to Know to Be Smarter, Safer, Wealthier, and Wiser*. Amherst, NY: Prometheus Books, 2015.

Harrison, Guy P. *Race and Reality: What Everyone Should Know About Our Biological Diversity*. Amherst, NY: Prometheus Books, 2010.

Hines, Terrence. *Pseudoscience and the Paranormal*. Amherst, NY: Prometheus Books, 2003.

Impey, Chris. *How It Ends: From You To the Universe*. New York, NY. W. W. Norton, 2011.

Kahneman, Daniel. *Thinking, Fast and Slow*. New York: Farrar, Straus, and Giroux, 2011.

Kaku, Michio. *The Future of the Mind: The Scientific Quest to Understand, Enhance, and Empower the Mind*. New York: Doubleday, 2014.

Kurzweil, Ray. *How to Create a Mind: The Secret of Human Thought Revealed*. New York: Penguin, 2012.

⸻. *The Singularity Is Near: When Humans Transcend Biology*. New York: Penguin, 2006.

Lane, Nick. *Life Ascending: The Ten Great Inventions of Evolution*. New York, NY: W. W. Norton, 2009.

Levy, Joel. *A Bee in a Cathedral*. Buffalo, NY: Firefly Books, 2011.

Lewin, Roger. *Human Evolution*, Malden, MA, USA: Blackwell, 2005.

Lynch, John, and Louise Barrett. *Walking with Cavemen*. New York: DK Publishing, 2003.

Margulis, Lynn; Sagan, Dorion. *What Is Life?* Berkley, CA: University of California Press, 2000.

McAuliffe, Kathleen. *This is Your Brain on Parasites: How Tiny Creatures Manipulate Our Behavior and Shape Society*. New York, NY: Houghman Mifflin Harcourt, 2016.

McRaney, David. *You Are Now Less Dumb: How to Conquer Mob Mentality, How to Buy Happiness, and All the Other Ways to Outsmart Yourself*. New York: Gotham. 2013.

———. *You Are Not So Smart: Why You Have Too Many Friends on Facebook, Why Your Memory Is Mostly Fiction, and 46 Other Ways You're Deluding Yourself*. New York: Gotham, 2012.

Medina, John. *Brain Rules: 12 Principles for Surviving and Thriving at Work, Home, and School*. Seattle: Pear, 2008.

Medina, John. *Brain Rules: 12 Principles for Surviving and Thriving at Work, Home, and School*. Seattle: Pear, 2008.

Merideth, Martin, *Born in Africa: The Quest for the Origins of Human Life*. New York, NY: Public Affairs, 2011.

Michael Shermer, *The Believing Brain: From Ghosts and Gods to Politics and Conspiracies—How We Construct Beliefs and Reinforce Them as Truths* (New York: Times Books, 2011), p. 278.

Montgomery, David R. and Bikle, Anne. *The Hidden Half of Nature: The Microbial Roots of Life and Health*. New York, NY: W.W. Norton, 2016.

Mooney, Chris, and Sheril Kirshenbaum. *Unscientific America: How Scientific Illiteracy Threatens Our Future*. New York: Basic Books, 2009.

Musser, George. *Spooky Action at a Distance: The Phenomenon That Reimagines Space and Time--and What It Means for Black Holes, the Big Bang, and Theories of Everything*. New York, NY: Scientific American, 2015.

New Scientist. *The Quantum World: The Disturbing Reality at the Heart of Reality*. Boston, MA: Nicholas Brealey Publishing, 2017.

Orzel, Chad. *How to Teach Quantum Physics to Your Dog*. New York, NY: Scribner, 2010.

Papagianni, Dimitra and Morse, Michael. *The Neanderthals Rediscovered: How Modern Science is Rewriting Their Story*. New York, NY: Thames and Hudson, 2013.

Plait, Philip. *Death From the Skies: The Science Behind the End of the World*. New York, NY: Penguin Books, 2008.

Prothero, Donald R. *The Story of Life in 25 Fossils*. New York, NY: Colombia University Press, 2015.

Pyne, Lydia. *Seven Skeletons: The Evolution of the World Most Famous Human Fossils*. New York, NY: Viking, 2016.

Ratey, John J. *Spark: The Revolutionary New Science of Exercise and the Brain*.

New York: Little, Brown, 2013.

Rees, Martin. *Our Final Hour*. New York, NY: Basic Books, 2003.

Relethford, John H.. *50 Great Myths of Evolution: Understanding Misconceptions about Our Origins*. West Sussex, UK: John Wiley and Sons, 2017.

Relethford, John H.; Mielke, James H.; Konigsberg, Lyle W. *Human Biological Variation*. Oxford, UK: Oxford University Press, 2011.

Renfrew, Colin. *Prehistory: The Making of the Human Mind*. New York, NY: Modern Library, 2008.

Reynolds, Gretchen. *The First 20 Minutes: Surprising Science Reveals How We Can Exercise Better, Train Smarter, Live Longer*. New York: Hudson Street, 2012.

Roossinck, Marilyn J. *Virus: An Illustrated Guide to 101 Incredible Microbes*. Princeton, NJ: Princeton University Press, 2016.

Rovelli, Carlo. *Reality Is Not What It Seems: The Journey to Quantum Gravity*. New York, NY: Riverhead, 2017.

Rovelli, Carlo. *Seven Brief Lessons on Physics*. New York, NY: Riverhead, 2016.

Sagan, Carl. *Billions & Billions: Thoughts on Life and Death at the Brink of the Millennium*. New York: Ballantine, 1998.

———. *The Demon-Haunted World: Science as a Candle in the Dark*. New York: Random House, 1995.

———. *Pale Blue Dot: A Vision of the Human Future in Space*. New York: Random House, 1994.

———. *The Varieties of Scientific Experience: A Personal View of the Search for God*. New York: Penguin, 2007.

Sagan, Carl, and Ann Druyan. Shadows of Forgotten Ancestors: A Search for Who We Are. New York: Random House, 1992.

Sapolsky, Robert M. *Behave: The Biology of Humans at Our Best and Worst*. New York, NY: Penguin Press, 2017.

Sawyer, G. J., and Victor Deak. *The Last Human: A Guide to Twenty-Two Species of Extinct Humans*. New Haven: Yale University Press, 2007.

Schick, Theodore, and Lewis Vaughn. *How to Think about Weird Things*. New York: McGraw-Hill, 2011.

Scharf, Caleb. *The Zoomable Universe*. New York, NY: Scientific American, 2017.

Seigel, Ethan. *Beyond the Galaxy: How Humanity Looked Beyond Our Milky Way and Discovered the Entire Universe*. Hackensack, NJ: World Scientific Publishing, 2015.

Shaw, Scott Richard. *Planet of the Bugs: Evolution and the Rise of Insects*. Chicago: University of Chicago Press, 2014.

Shermer, Michael. The Believing Brain: From Ghosts and Gods to Politics and Conspiracies—How We Construct Beliefs and Reinforce Them as Truths. New York: Times Books, 2011

Shubin, Neil. *Your Inner Fish: A Journey Into the 3.5 Billion-Year History of the*

Human Body. New York, NY: Pantheon, 2008.

———. *The Universe Within: The Deep History of the Human Body*. New York, NY: First Vintage Books, 2013.

Smith, Cameron and Sullivan, Charles. *The Top 10 Myths about Evolution*. Amherst, NY: Prometheus Books, 2007.

Smith, Cameron. *The Fact of Evolution*. Amherst, NY: Prometheus Books, 2011.

Smith, David Livingston. *Less Than Human: Why We Demean, Enslave, and Exterminate Others*. New York, NY: St. Martin's Press, 2011.

Smith, David Livingston. *The Most Dangerous Animal: Human Nature and the Origins of War*. New York, NY: St. Martin's Press, 2007.

Smith, Jonathan C. *Pseudoscience and Extraordinary Claims of the Paranormal: A Critical Thinker's Toolkit*. West Sussex, UK: Wiley-Blackwell, 2010.

Stanford, Craig. Significant Others: The Ape-Human Continuum and the Quest for Human Nature. New York, NY: Basic Books, 2001.

Stringer, Chris, and Peter Andrews. *The Complete World of Human Evolution*. New York: Thames and Hudson, 2005.

Stringer, Chris. *Lone Survivors: How We Came to Be the Only Humans on Earth*. New York, NY: Times Books/Henry Holt, 2012.

Suzuki, Wendy, with Fitzpatrick, Billie. *Healthy Brain, Happy Life: A Personal Program to Activate Your Brain and Do Everything Better*. New York: Harper Collins, 2015.

Tattersall, Ian. *Becoming Human: Evolution and Human Uniqueness*. New York: Harcourt Brace, 1998.

———. *Extinct Humans*. New York: Basic Books, 2001.

———. *The Human Odyssey: Four Million Years of Human Evolution*. New York: Prentice Hall, 1993.

———. *Masters of the Planet: The Search for Our Human Origins*. New York: Palgrave Macmillan, 2012.

Tegmark, Max. *Life 3.0: Being Human in the Age of Artificial Intelligence*. New York, NY: Alfred A. Knopf, 2017.

Torres, Phil. *The End: What Science and Religion Tell Us about the Apocalypse*. Durham, NC: Pitchstone, 2016.

———. *Morality, Foresight, and Human Flourishing: An Introduction to Existential Risks*. Durham, North Carolina: Pitchstone Publishing, 2017.

Tudge, Colin. *The Tree: A Natural History of What Trees Are, How They Live, and Why They Matter*. New York, NY: Three Rivers Press, 2005.

Tyson, Neil DeGrasse, and Donald Goldsmith. Origins: Fourteen Billion Years of Cosmic Evolution. New York: W. W. Norton, 2004.

Tyson, Neil DeGrasse; Goldsmith, Donald. *Origins: Fourteen Billion Years of Cosmic Evolution*. New York: W. W. Norton, 2004.

Tyson, Neil DeGrasse; Strauss, Michael A.; Gott, J. Richard. *Welcome to the Universe: An Astrophysical Tour*. Princeton, NJ: Princeton University

Press, 2016.

Walter, Chip. *Last Ape Standing: The Seven-Million-Year Old Story of How and Why We Survived*. New York, NY: Walker and Company, 2013.

Ward, Peter and Kirschvink, Joe. *A New History of Life: The Radical New Discoveries about the Origins and Evolution of Life on Earth*. New York, NY: Bloomsbury Press, 2015.

Ward, Peter. *Life as We Do Not Know It*. New York, NY: Viking, 2005.

Weinberg, Steven. *Facing Up: Science and Its Cultural Adversaries*. Cambridge, MA: Harvard University Press, 2003.

———. *Lake Views: This World and the Universe*. Cambridge, MA: Belknap Press of Harvard University Press, 2010.

Wells, Spencer. *Pandora's Seed: The Unforeseen Cost of Civilization*. New York: Random House, 2010.

Wilson E. O. *The Future of Life*. New York, NY: Alfred A. Knopf, 2002.

———. *The Social Conquest of Earth*. New York: Liveright Publishing, 2012.

Wilson, Edward O. The Meaning of Existence. New York: W. W. Norton, 2014.

Wohlleben, Peter. *The Hidden Life of Trees: What They Feel, How They Communicate*. Vancouver, BC: Greystone Books, 2016.

Wrangham, Richard. *Catching Fire: How Cooking Made Us Human*. New York: Basic Books, 2009.

Young, Ed. *I Contain Multitudes: The Microbes Within Us and a Grander View of Life*. New York, NY. HarperCollins, 2016.

Zimmer, Carl. *Evolution: The Triumph of an Idea*. New York: Harper Perennial, 2006.

NOTES

INTRODUCTION

1. National Science Board, "Science and Engineering Indicators", https://www.nsf.gov/statistics/2016/nsb20161/#/report/chapter-7/public-knowledge-about-s-t, (accessed November 9, 2017).

2. *Michigan News*, "U.S. public's knowledge of science: getting better but a long way to go", Feb 15, 2011, http://ns.umich.edu/new/releases/8265-u-s-public-s-knowledge-of-science-getting-better-but-a-long-way-to-go

3. *Michigan News*, Ibid.

4. *Michigan News*, Ibid.

5. AP-GfK, "The AP-GfK Poll", March, 2014, p. 2. http://ap-gfkpoll.com/main/wp-content/uploads/2014/04/AP-GfK-March-2014-Poll-Topline-Final_SCIENCE.pdf (accessed November 9, 2017).

6. AP-GfK, Ibid.

CHAPTER 1

7. Roesch LF, Fulthorpe RR, Riva A, Casella G, Hadwin AK, Kent AD, Daroub SH, Camargo FA, Farmerie WG, Triplett EW, "Pyrosequencing enumerates and contrasts soil microbial diversity," *ISME*, August, 2007; 1(4):283-90. https://www.ncbi.nlm.nih.gov/pubmed/18043639 (accessed September 15, 2017).

8. Camilo Mora, Derek P. Tittensor, Sina Adl, Alastair G. B. Simpson, Boris Worm, "How Many Species Are There on Earth and in the Ocean?" PLOS Biology, August 23, 2011 https://doi.org/10.1371/journal.pbio.1001127 (accessed November 9, 2017).

9. John Cryan, interview with the author, January 27, 2018.

10. MediaCollege.com, "How to Write a News Story," http://www.mediacollege.com/journalism/news/write-stories.html (accessed September 15, 2017).

11. National Center for Science Education, "Americans' Scientific Knowledge and Beliefs about Human Evolution in the Year of Darwin," https://ncse.com/library-resource/americans-scientific-knowledge-beliefs-human-evolution-

year (accessed September 5, 2017).

12. Samantha Grossman, "1 in 4 Americans Apparently Unaware the Earth Orbits the Sun," Time, Feb 16, 2014 http://time.com/7809/1-in-4-americans-thinks-sun-orbits-earth/ (accessed September 5, 2017).

13. Cary Funk, Sara Kehaulani Goo, "A Look at What the Public Knows and Does Not Know About Science," Pew Research Center, September 10, 2015, http://www.pewinternet.org/2015/09/10/what-the-public-knows-and-does-not-know-about-science/ (accessed November 13, 2017).

14. Donald Lowe, interview with the author, January 25, 2018.

15. Funk and Goo, Ibid.

16. NASA Jet Propulsion Laboratory, "NASA Office to Coordinate Asteroid Detection, Hazard Mitigation," January 7, 2016, https://www.jpl.nasa.gov/news/news.php?feature=4816 (accessed October 25, 2017).

17. Jeff Daniels, "The Navy's put down a 'significant bet' on the $13 billion USS Gerald R Ford, which some say is a risky gamble," CNBC, July 22, 2017, https://www.cnbc.com/2017/07/22/ford-carrier-emblematic-of-navys-struggle-with-technology-costs.html (accessed October 25, 2017).

18. Sean Rossman, "Americans are spending more than ever on plastic surgery," USA TODAY, April 12, 2017, https://www.usatoday.com/story/news/nation-now/2017/04/12/americans-spending-more-than-ever-plastic-surgery/100365258/ (accessed October 25, 2017).

19. Beverly Peterson Stearns and Stephen C. Stearns, Watching from the Edge of Extinction, Yale University Press, 2000, p. x.

Richard Dawkins, Science in the Soul: Selected Writings of a Passionate Rationalist, (Bantam Press, London), p. 71.

20. István Horváth, Zsolt Bagoly, Jon Hakkila, and L. V. Tóth, "New data support the existence of the Hercules-Corona Borealis Great Wall," Astronomy and Astrophysics, Volume 584, December 2015, https://www.aanda.org/articles/aa/abs/2015/12/aa24829-14/aa24829-14.html (accessed September 10, 2017).

21. National Science Board, "Science and Engineering Indicators," https://www.nsf.gov/statistics/2016/nsb20161/#/report/chapter-7/public-knowledge-about-s-t, (accessed November 9, 2017).

22. Chapman University, "Paranormal America 2017: Chapman University Survey of American Fears 2017," October 11, 2017, https://blogs.chapman.edu/wilkinson/2017/10/11/paranormal-america-2017/ (accessed November 13, 2017).

23. Joe Rogan Experience, podcast #938, March 27, 2017, https://www.youtube.com/watch?v=bDhHK8nk_V0 (accessed September 14, 2017).

24. Edward O. Wilson, The Social Conquest of Earth (New York: W. W. Norton, 2012), p. 7.

CHAPTER 2

25. Stephen Hawking, "The Beginning of Time," http://www.hawking.org.uk/, http://www.hawking.org.uk/the-beginning-of-time.html (accessed January 17, 2018).

Scientific American, "According to the big bang theory, all the matter in the universe erupted from a singularity. Why didn't all this matter--cheek by jowl as it was--immediately collapse into a black hole?" September 22, 2003 (accessed January 17, 2018).

26. Brian Koberlein, interview with the author. December 15, 2017.

27. Brian Greene, *The Fabric of the Cosmos: Space, Time, and the Texture of Reality*, (New York: First Vintage Books Edition, 2005), p. 272.

28. Sean Carroll, *The Big Picture: On the Origins of Life, Meaning, and the Universe Itself*, (New York, NY: Dutton, 2016), p. 51.

29. Michael A. Strauss, *Welcome to the Universe: An Astrophysical Tour*, (Princeton, NJ: Princeton University Press, 2016), p. 227.

30. NASA, "Tests of Big Bang: The CMB," https://wmap.gsfc.nasa.gov/universe/bb_tests_cmb.html (accessed December 1, 2017).

31. NASA, "Dark Energy, Dark Matter," https://science.nasa.gov/astrophysics/focus-areas/what-is-dark-energy (accessed December 10, 2017).

32. Ibid.

33. NASA, "WMAP Produces New Results," https://map.gsfc.nasa.gov/news/ (accessed December 10, 2017).

34. C. R. Lawrence, "Planck 2015 Results", Astrophysics Subcommittee, NASA, p. 29, https://smd-prod.s3.amazonaws.com/science-green/s3fs-public/mnt/medialibrary/2015/04/08/CRL_APS_2015-03-18_compressed2.pdf (accessed December 10, 2017). And European Space Agency, "Planck Overview," http://www.esa.int/Our_Activities/Space_Science/Planck_overview (accessed December 10, 2017).

35. Marcelo Gleiser, "What Does An Expanding Universe Really Mean?" Cosmos and Culture, NPR, May 24, 2017, https://www.npr.org/sections/13.7/2017/05/24/529675773/what-does-an-expanding-universe-really-mean (accessed January 18, 2018).

36. Walter Isaacson, *Einstein: His Life and Universe*, New York, NY: Simon and Shuster, 2007), pp. 254-5, 353-56.

37. Ibid, pp. 356.

Space Telescope Science Institute, "What Is Dark Energy?" http://hubblesite.org/hubble_discoveries/dark_energy/de-what_is_dark_energy.php (accessed January 18, 2018).

Space Telescope Science Institute, "Did Einstein Predict Dark Energy?" http://hubblesite.org/hubble_discoveries/dark_energy/de-did_einstein_predict.php (accessed January 18, 2018).

38. A. Bonanno, H. Schlattl, and L. Paternò, "The age of the Sun and the relativistic corrections in the EOS," *Astronomy and Astrophysics*, V. 390, Number, 3, August II 2002, pp. 1115 – 1118, https://doi.org/10.1051/0004-6361:20020749 (accessed December 4, 2017).

39. Holly Zell, Jennifer Rumburg, NASA, "Heliophysics Fun Facts," https://www.nasa.gov/mission_pages/sunearth/overview/Helio-facts.html#helio (accessed December 4, 2017).

40. Ibid (Zell and Rumburg).

41. Encyclopaedia Britannica, "Sun Worship," https://www.britannica.com/topic/sun-worship (accessed December 4, 2017).

42. Maria Temming, "How Many Stars are There in the Universe?", July 15, 2014 http://www.skyandtelescope.com/astronomy-resources/how-many-stars-are-there/ (accessed December 24, 2017).

43. Ibid, Holly Zell, Jennifer Rumburg. And, NASA, "Our Star the Sun," https://sohowww.nascom.nasa.gov/classroom/classroom.html (accessed December 7, 2017).

44. Ibid, Beverly Peterson Stearns and Stephen C. Stearns, 2000.

45. NASA, "What is the speed of the Earth's rotation?" <INSERT>Image Education Center, https://<INSERT>Image.gsfc.nasa.gov/poetry/ask/a10840.html (accessed December 1, 2017).

46. Live Science, "How Fast Does Earth Move?" November 27, 2012, https://www.livescience.com/32294-how-fast-does-earth-move.html (accessed December 1, 2017).

47. Rhett Herman, *Scientific American*, "How Fast is the Earth Moving?," https://www.scientificamerican.com/article/how-fast-is-the-earth-mov/ (accessed December 1, 2017).

48. NASA, "Earth in Depth," https://solarsystem.nasa.gov/planets/earth/indepth (accessed December 4, 2017). And Eugene C. Robertson, "Interior of the Earth," USGS, https://pubs.usgs.gov/gip/interior/ (accessed December 6, 2017).

49. G. Brent Dalrymple, "The age of the Earth in the twentieth century: a problem (mostly) solved," Geological Society, London, Special Publications, 190, 205-221, 1 January 2001, https://doi.org/10.1144/GSL.SP.2001.190.01.14 (accessed December 3, 2017).

50. Carl Sagan, *Cosmos*, episode one (near the end), and *The Dragons of Eden*, (New York, NY: Random House, 1977), p. 15, and Therese Puyau Blanchard and Andy Fraknoi, "Cosmic Calendar," Astronomical Society of the Pacific, 2010, https://astrosociety.org/edu/astro/act2/H2_Cosmic_Calendar.pdf (accessed December 4, 2017).

51. Ibid, Dalrymple, 2001.

52. G. Brent Dalrymple, *The Age of the Earth*, (Palo Alto, California: *Stanford University Press, 1994)*.

53. G. Brent Dalrymple, "The Ages of the Earth, Solar System, Galaxies,

and Universe," in *Scientists Confront Intelligent Design and Creationism,*" (New York, NY: W.W. Norton, 2007), pp. 151-152.

54. Gary Nichols, *Sedimentology and Stratigraphy* (New York, NY: John Wiley & Sons, 2009), pp. 325–327

55. Samuel A. Bowring, Ian S. Williams, "Priscoan (4.00–4.03 Ga) orthogneisses from northwestern Canada," *Contributions to Mineralogy and Petrology*, January 1999, Volume 134, Issue 1, pp 3–16.

56. US Geological Survey, "Age of the Earth," July 9, 2007, https://pubs. usgs.gov/gip/geotime/age.html (accessed October 29, 2017).

57. National Center for Science Education, "Americans' Scientific Knowledge and Beliefs about Human Evolution in the Year of Darwin", "https://ncse. com/library-resource/americans-scientific-knowledge-beliefs-human-evolution-year (accessed September 5, 2017).

Tia Ghose, "4 in 10 Americans Believe God Created Earth 10,000 Years Ago," Live Science, June 5, 2014, https://www.livescience.com/46123-many-americans-creationists.html (accessed January 18, 2018).

Frank Newport, "In U.S., 42% Believe Creationist View of Human Origins," Gallup News, June 2, 2014, http://news.gallup.com/poll/170822/believe-creationist-view-human-origins.aspx (accessed January 18, 2018).

David Masci, "For Darwin Day, 6 facts about the evolution debate," Pew Research Center, February 10, 2017, http://www.pewresearch.org/fact-tank/2017/02/10/darwin-day/ (accessed December 6, 2017).

58. Old Earth Ministries, "Can You Be a Christian and Believe in an Old Earth?" http://www.oldearth.org/question.htm (accessed January 18, 2018).

59. Nick Squires, "Pope Francis says Big Bang theory and evolution 'compatible with divine Creator'," *The Telegraph*, October 28, 2014, http://www. telegraph.co.uk/news/worldnews/the-pope/11192802/Pope-Francis-says-Big-Bang-theory-and-evolution-compatible-with-divine-Creator.html   (accessed January 18, 2018).

60. Ibid, Dalrymple, 2007, p. 161.

61. NASA/Jet Propulsion Laboratory, "Voyager: Fast Facts," https:// voyager.jpl.nasa.gov/frequently-asked-questions/fast-facts/   (accessed December 6, 2017).

62. Eugene C. Robertson, "Interior of the Earth," USGS, https://pubs.usgs. gov/gip/interior/ (accessed December 6, 2017).

63. Physics.org, "What Causes the Earth's Magnetic Field?," http://www. physics.org/article-questions.asp?id=64 (accessed November 9, 2017).

64. Lee Pullen, "Plate Tectonics Could be Essential for Life," *Astrobiology Magazine*, Feb 19, 2009, https://www.astrobio.net/news-exclusive/plate-tectonics-could-be-essential-for-life/ (accessed December 6, 2017).

65. Ibid.

66. Leonardo Calle et al, "Effects of tidal periodicities and diurnal foraging constraints on the density of foraging wading birds," *The Auk: Ornithological*

Advances, May, 2016 DOI: 10.1642/AUK-15-234.1 (accessed December 7, 2017). And, Je' Czaja, "What Is the Importance of the Intertidal Zone?" *Sciencing*, https://sciencing.com/importance-intertidal-zone-6856404.html   (accessed December 7, 2017).

67. Richard Lathe, "Fast tidal cycling and the origin of life," *Icarus*, Volume 168, Issue 1, March 2004, Pages 18-22.

68. Richard Lathe, interview with the author, December 12, 2017.

69. Melanie Barboni, et al, "Early formation of the Moon 4.51 billion years ago," *Science Advances*, 11 January, 2017: Vol. 3, no. 1, e1602365 DOI: 10.1126/sciadv.1602365 (accessed December 7, 2017).

70. NASA, "Earth's Moon: In Depth," https://solarsystem.nasa.gov/planets/moon/indepth (accessed November 1, 2017). And, NASA, "Earth's Moon: 10 Need-To-Know Things," https://solarsystem.nasa.gov/planets/moon/needtoknow (accessed November 1, 2017).

71. David Morse, "Lunar Data Support Idea That Collision Split Earth, Moon," NASA, March 16, 1999, https://www.nasa.gov/centers/ames/news/releases/1999/99_18AR.html (accessed December 6, 2017).

And Robin M. Canup, "Forming a Moon with an Earth-like Composition via a Giant Impact," *Science*, 23 November 2012, Vol. 338, Issue 6110, pp. 1052-1055 DOI: 10.1126/science.1226073 (accessed December 6, 2017).

72. Nola Taylor Redd, "How Was the Moon Formed," Space.com, November 15, 2017, https://www.space.com/19275-moon-formation.html   (accessed December 6, 2017).

73. Neil DeGrasse Tyson and Donald Goldsmith, *Origins: Fourteen Billion Years of Cosmic Evolution*, (New York, NY: W. W. Norton, 2004), p. 192.

CHAPTER 3

74. Kenneth J. Locey, Jay T. Lennona, "Scaling laws predict global microbial diversity", Proceedings of the National Academy of Sciences, 5970–5975, vol. 113 no. 21, May 24, 2016, doi: 10.1073/pnas.1521291113 (accessed September 27, 2017).

75. Thomas Jefferson National Accelerator Facility, "Questions and Answers", https://education.jlab.org/qa/mathatom_04.html (accessed October 1, 2017).

76. Graham Lawton, *The Origin of (Almost) Everything*, (London, UK: John Murray, 2016), pp. 24-25.

77. Thomas Jefferson National Accelerator Facility, "Questions and Answers", https://education.jlab.org/qa/mathatom_04.html (accessed October 1, 2017).

78. U.S. Geological Survey's (USGS) Water Science School "The Water In

Notes   275

You", 2016, https://water.usgs.gov/edu/propertyyou.html (accessed October 2, 2017).

79. Ron Sender, Shai Fuchs, Ron Milo "Revised Estimates for the Number of Human and Bacteria Cells in the Body", PLOS, August 19, 2016, https://doi.org/10.1371/journal.pbio.1002533 (accessed September 30, 2017).

80. Ron Sender, Shai Fuchs, Ron Milo "Revised Estimates for the Number of Human and Bacteria Cells in the Body", *PLOS*, August 19, 2016, https://doi.org/10.1371/journal.pbio.1002533 (accessed September 30, 2017).

81. Laura Geggel, "How Much Blood Is in the Human Body?", Live Science, March 3, 2016, https://www.livescience.com/32213-how-much-blood-is-in-the-human-body.html (accessed October 31, 2017).

82. "Muscular System", InnerBody, http://www.innerbody.com/Image/musfov.html (accessed October 31, 2017).

83. Ron Sender, Shai Fuchs, Ron Milo "Revised Estimates for the Number of Human and Bacteria Cells in the Body", *PLOS*, August 19, 2016, https://doi.org/10.1371/journal.pbio.1002533 (accessed September 30, 2017).

84. Erin Allday, "100 trillion good bacteria call human body home", *SFGate*, July 5, 2012, http://www.sfgate.com/health/article/100-trillion-good-bacteria-call-human-body-home-3683153.php (accessed October 2, 2017).

85. David Enard, Le Cai, Carina Gwennap, Dmitri A Petrov, "Viruses are a dominant driver of protein adaptation in mammals", eLife, 2016; 5, DOI: 10.7554/eLife.12469 (accessed November 3, 2017).

86. Belshaw R, Pereira V, Katzourakis A, Talbot G, Paces J, Burt A, Tristem M (April 2004). "Long-term reinfection of the human genome by endogenous retroviruses". *Proceedings of the National Academy of Sciences*. 101 (14): 4894–9. doi:10.1073/pnas.0307800101 (accessed November 3, 2017).

Julia Halo Wildschutte, Zachary H. Williams, Meagan Montesion, Ravi P. Subramanian, Jeffrey M. Kidd, John M. Coffin, "Discovery of unfixed endogenous retrovirus insertions in diverse human populations", *Proceedings of the National Academy of Sciences*, 2016; 201602336 DOI: 10.1073/pnas.1602336113 (accessed November 3, 2017).

E. B. Chuong, N. C. Elde, C. Feschotte. "Regulatory evolution of innate immunity through co-option of endogenous retroviruses", *Science*, 2016; 351 (6277): 1083 DOI: 10.1126/science.aad5497 (accessed November 3, 2017).

87. Carl Zimmer, "Ancient Viruses Are Buried in Your DNA", New York Times, October 4, 2017, https://www.nytimes.com/2017/10/04/science/ancient-viruses-dna-genome.html (accessed October 4, 2017).

88. James F. Meadow, et al. "Humans Differ in Their Personal Microbial Cloud." Ed. Valeria Souza. PeerJ 3 (2015): e1258. PMC. https://www.ncbi.nlm.nih.gov/pmc/articles/PMC4582947/ (accessed December 30, 2017).

89. Rasnik K. Singh, Hsin-Wen Chang, et al., "Influence of diet on the gut microbiome and implications for human health," *Journal of translational medicine*," April 8, 2017; 15: 73, doi: 10.1186/s12967-017-1175-y (accessed January

17, 2018).

Science, "Your Microbes, Your Health," 20 Dec 2013: Vol. 342, Issue 6165, pp. 1440-1441 DOI: 10.1126/science.342.6165.1440-b (accessed January 17, 2018).

90. Mark Kowarsky, Joan Camunas-Soler, Michael Kertesz, Iwijn De Vlaminck, Winston Koh, Wenying Pan, Lance Martin, Norma F. Neff, Jennifer Okamoto, Ronald J. Wong, Sandhya Kharbanda, Yasser El-Sayed, Yair Blumenfeld, David K. Stevenson, Gary M. Shaw, Nathan D. Wolfe, Stephen R. Quake. "Numerous uncharacterized and highly divergent microbes which colonize humans are revealed by circulating cell-free DNA", *Proceedings of the National Academy of Sciences*, 2017; 201707009 DOI: 10.1073/pnas.1707009114 (accessed September 16, 2017).

Nathan Collins, "Stanford study indicates that more than 99 percent of the microbes inside us are unknown to science" *Stanford News*, August 22, 2017, http://news.stanford.edu/2017/08/22/nearly-microbes-inside-us-unknown-science/ (accessed October 3, 2017).

91. Marc Ereshefsky, "Some Problems with the Linnaean Hierarchy", *Philosophy of Science*, Vol. 61, No. 2 (Jun., 1994), pp. 186-205.

92. Marta Paterlini, "There shall be order. The legacy of Linnaeus in the age of molecular biology," *EMBO Reports*, 2007 Sep; 8(9): 814–816. doi: 10.1038/sj.embor.7401061 (accessed January 20, 2018).

93. Zhongwei Guo, Lin Zhang, Yiming Li, "Increased Dependence of Humans on Ecosystem Services and Biodiversity", *PLOS One*, October 1, 2010, https://doi.org/10.1371/journal.pone.0013113 (accessed September 30, 2017).

94. David George Haskell, *The Forest Unseen: Eight Years Watch in Nature*, (New York; Penguin Books, 2012), p. 245.

95. Spencer Wells, *Pandora's Seed: The Unforeseen Cost of Civilization*, (New York: Random House, 2010), pp. 115-121.

96. David G. Pearson, Tony Craig, "The great outdoors? Exploring the mental health benefits of natural environments", *Frontiers in Psychology*, 2014; 5: 1178, doi: 10.3389/fpsyg.2014.01178 (accessed September 22, 2017).

97. Frances E. Kuo, William C. Sullivan, "Aggression and Violence in the Inner City: Effects of Environment via Mental Fatigue", *Environment and Behavior*, July 1, 2001, Volume: 33 issue: 4, page(s): 543-571, https://doi.org/10.1177/00139160121973124 (accessed September 17, 2017).

98. Gregory N. Bratman, J. Paul Hamilton, "Nature experience reduces rumination and subgenual prefrontal cortex activation," *Proceedings of the National Academy of Sciences*, vol. 112 no. 28, doi: 10.1073/pnas.1510459112 (accessed January 20, 2018).

99. Ibid.

100. Simone Kühn, Sandra Düzel, Peter Eibich, Christian Krekel, Henry Wüstemann, Jens Kolbe, Johan Martensson, Jan Goebel, Jürgen Gallinat, Gert G. Wagner, Ulman Lindenberger, "Associations between geographical proper-

ties and brain structure", *Scientific Reports,* 7, Article number: 11920 (2017) doi:10.1038/s41598-017-12046-7 (accessed October 20, 2017).

Florence Williams, *The Nature Fix: Why Nature Makes Us Happier, Healthier, and more Creative,* (New York: W.W. Norton, 2017).

101. Beyer, K. M., et al. "Exposure to neighborhood green space and mental health: evidence from the survey of the health of Wisconsin", *International Journal Environ Res Public Health,* 11, 3453–3472, 2014, https://doi.org/10.3390/ijerph110303453 (accessed October 20, 2017).

102. Maas, J., Verheij, R. A., Groenewegen, P. P., de Vries, S. & Spreeuwenberg, P., "Green space, urbanity, and health: how strong is the relation?" *Journal of Epidemiology and Community Health* 60, 587–592, 2006, https://doi.org/10.1136/jech.2005.043125 (accessed October 20, 2017).

103. Geoffrey H. Donovan, et al., "The Relationship Between Trees and Human Health: Evidence from the Spread of the Emerald Ash Borer", *American Journal of Preventive Medicine,* February 2013 ,Volume 44, Issue 2, Pages 139–145, DOI: http://dx.doi.org/10.1016/j.amepre.2012.09.066 (accessed October 20, 2017).

104. Takano, T., Nakamura, K. & Watanabe, M. "Urban residential environments and senior citizens' longevity in megacity areas: the importance of walkable green spaces", *Journal of Epidemiology and Community Health,* 56, 2002, pp. 913–918.

Mitchell, R. & Popham, F., "Effect of exposure to natural environment on health inequalities: an observational population study", *Lancet,* 372, 2008, https://doi.org/10.1016/S0140-6736(08)61689-X (accessed October 4, 2017).

105. Nick Patterson, Daniel J. Richter, Sante Gnerre, Eric S. Lander, David Reich, "Genetic evidence for complex speciation of humans and chimpanzees", *Nature* 441, 1103-1108 (29 June 2006) | doi:10.1038/nature04789 (accessed October 29, 2017).

106. Charles Q. Choi, "Fossil Reveals What Last Common Ancestor of Humans and Apes Looked Like", *Scientific American,* August 10, 2017, https://www.scientificamerican.com/article/fossil-reveals-what-last-common-ancestor-of-humans-and-apes-looked-liked/ (accessed October 31, 2017).

Arnason U1, Gullberg A, Janke A., "Molecular timing of primate divergences as estimated by two nonprimate calibration points", *Journal of Molecular Evolution,* J Mol Evol. 1998 Dec;47(6):718-27.

https://www.ncbi.nlm.nih.gov/pubmed/9847414 (accessed September 29, 2017).

Priya Moorjani, Carlos Eduardo G. Amorim, Peter F. Arndt, and Molly Przeworskia, "Variation in the molecular clock of primates", Proceedings of the National Academy of Sciences of the U S A, 2016 Sep 20; 113(38): 10607–10612, doi: 10.1073/pnas.1600374113 (accessed October 29, 2017).

107. Colin Barras, "Just how are we related to our chimp cousins?", *New Scientist,* 16 March 2016, https://www.newscientist.com/article/2081012-just-

how-are-we-related-to-our-chimp-cousins/ (accessed September 29, 2017).

108. Kate Wong, "Tiny Genetic Differences between Humans and Other Primates Pervade the Genome

Genome comparisons reveal the DNA that distinguishes Homo sapiens from its kin", Scientific American, September 1, 2014, https://www.scientificamerican.com/article/tiny-genetic-differences-between-humans-and-other-primates-pervade-the-genome/ (accessed January 2, 2018).

109. Ann Gibbons, "Bonobos Join Chimps as Closest Human Relatives", *Science*, June 13, 2012, http://www.sciencemag.org/news/2012/06/bonobos-join-chimps-closest-human-relatives (accessed January 2, 2018).

110. The Orangutan Project, "Orangutan Facts", https://www.theorangutanproject.org/about-orangutans/orangutan-facts/ (accessed January 2, 2018).

111. Kay Prüfer, Kasper Munch, Ines Hellmann, et al., "The bonobo genome compared with the chimpanzee and human genomes", *Nature*, 486, 527–531 (28 June 2012) doi:10.1038/nature11128 (accessed August 7, 2017).

112. Ibid.

113. *"Homo habilis"*, Smithsonian National Museum of Natural History, http://humanorigins.si.edu/evidence/human-fossils/species/homo-habilis (accessed October 29, 2017).

Leakey, L.S.B., Tobias, P.V., Napier, J.R., "A new species of the genus Homo from Olduvai Gorge", *Nature* 202, 1964, 7-9. http://www.tarha.ulpgc.es/leakey_1964.pdf (accessed October 29, 2017).

114. Antón, S. C. (2003), Natural history of *Homo erectus. American Journal of Physical Anthropology*, 122: 126–170. doi:10.1002/ajpa.10399 (accessed October 29, 2017).

"Homo erectus", Smithsonian National Museum of Natural History, http://humanorigins.si.edu/evidence/human-fossils/species/homo-erectus (accessed October 29, 2017).

115. Lee Berger and John Hawks, *Almost Human: The Astonishing Tale of Homo naledi and the Discovery That Changed Our Human Story*, (Des Moines, IA: National Geographic, 2017).

Lee R. Berger, John Hawks, et al., "Homo naledi, a new species of the genus Homo from the Dinaledi Chamber, South Africa", eLife 2015;4:e09560,eLife 2015;4:e09560 eLife 2015;4:e09560doi: 10.7554/eLife.09560 (accessed October 30, 2017).

Chris Stringer, "Human Evolution: The many mysteries of *Homo naledi"*, eLife 2015;4:e10627, eLife 2015;4:e10627 doi: 10.7554/eLife.10627 (accessed October 30, 2017).

Paul HGM Dirks, et al., "The age of Homo naledi and associated sediments in the Rising Star Cave, South Africa", eLife 2017;6:e24231 doi: 10.7554/eLife.24231 eLife 2017;6:e24231 doi: 10.7554/eLife.24231 (accessed October 30, 2017).

. Michael Greshko, "Did This Mysterious Ape-Human Once Live Alongside

Our Ancestors?", National Geographic, May 9, 2017, https://news.nationalgeographic.com/2017/05/homo-naledi-human-evolution-science/ (accessed October 30, 2017).

116. *"Homo heidelbergensis"*, Smithsonian National Museum of Natural History, http://humanorigins.si.edu/evidence/human-fossils/species/homo-heidelbergensis (accessed October 29, 2017).

Mounier A, Marchal F, Condemi S., "Is Homo heidelbergensis a distinct species? New insight on the Mauer mandible", *Journal of Human Evolution*, 2009 March, 56(3):219-46. doi: 10.1016/j.jhevol.2008.12.006 (accessed October 29, 2017).

117. Bridget Alex, "Meet the Denisovans, *Discover*, November 4, 2016, http://discovermagazine.com/2016/dec/meet-the-denisovans (accessed October 30, 2017).

Carl Zimmer (22 December 2010). "Denisovans Were Neanderthals' Cousins, DNA Analysis Reveals". *NYTimes.com.* "Denisovans Were Neanderthals' Cousins, DNA Analysis Reveals" (accessed October 30, 2017).

Katherine Harmon, "New DNA Analysis Shows Ancient Humans Interbred with Denisovans", *Scientific American*, August 30, 2012, https://www.scientificamerican.com/article/denisovan-genome/ (accessed October 30, 2017).

118. *"Homo floresiensis"*, Smithsonian National Museum of Natural History, http://humanorigins.si.edu/evidence/human-fossils/species/homo-floresiensis (accessed October 29, 2017).

Australian National Museum, "Origins of Indonesian 'hobbits' finally revealed", 21 April 2017, http://www.anu.edu.au/news/all-news/origins-of-indonesian-%E2%80%98hobbits%E2%80%99-finally-revealed (accessed October 29, 2017).

Mike Morwood, Penny van Oosterzee, *A New Human: The Startling Discovery and Strange Story of the "Hobbits" of Flores, Indonesia*, (New York, NY: Smithsonian, 2007).

119. *"Homo neanderthalensis"* – The Neanderthals", Australian Museum, https://australianmuseum.net.au/homo-neanderthalensis (accessed October 29, 2017).

Dimitra Papagianni and Michael Morse, *The Neanderthals Rediscovered: How Modern Science Is Rewriting Their Story*, (New York, NY: Thames & Hudson, 2013).

Ian Tattersall, *The Last Neanderthal: The Rise, Success, and Mysterious Extinction of Our Closest Human Relatives*, New York, NY: Macmillan General Reference, 1996).

The Genographic Project, "Why Am I a Neanderthal", National Geographic, https://genographic.nationalgeographic.com/neanderthal/ (accessed October 30, 2017).

120. Max Roser, "Human Height", OurWorldInData.org, https://ourworldindata.org/human-height/ (accessed October 29, 2017).

121. Walpole, S.C., Prieto-Merino, D., Edwards, P. et al. "The weight of nations: an estimation of adult human biomass", *BMC Public Health*, (2012) 12: 439. https://link.springer.com/article/10.1186/1471-2458-12-439 (accessed October 29, 2017). Walpole, S.C., Prieto-Merino, D., Edwards, P. et al. BMC Public Health (2012) 12: 439. Walpole, S.C., Prieto-Merino, D., Edwards, P. et al. BMC Public Health (2012) 12: 439.Walpole, S.C., Prieto-Merino, D., Edwards, P. et al. BMC Public Health (2012) 12: 439.

122. Chris Stringer, *Lone Survivors: How We Came to Be the Only Humans on Earth*, (New York, NY: Times Books/Henry Holt, 2012).

Ian Tattersall, *Masters of the Planet: The Search for Our Human Origins*, (New York, NY: St. Martin's Griffin, 2013).

Lydia Pyne, *Seven Skeletons: The Evolution of the World's Most Famous Human Fossils*, (New York, NY: Viking, 2016).

123. Thomas Sutikna, Matthew W. Tocheri, Michael J. Morwood, E. Wahyu Saptomo, Jatmiko, Rokus Due Awe, Sri Wasisto, Kira E. Westaway, Maxime Aubert, Bo Li, Jian-xin Zhao, Michael Storey, Brent V. Alloway, Mike W. Morley, Hanneke J. M. Meijer, Gerrit D. van den Bergh, Rainer Grün, Anthony Dosseto, Adam Brumm, William L. Jungers & Richard G. Roberts, "Revised stratigraphy and chronology for Homo floresiensis at Liang Bua in Indonesia", *Nature*, 532, 366–369 (21 April 2016) doi:10.1038/nature17179 (accessed September 11, 2017).

124. Tom Higham, Katerina Douka, Rachel Wood, Christopher Bronk Ramsey, Fiona Brock, Laura Basell, Marta Camps, Alvaro Arrizabalaga, Javier Baena, Cecillio Barroso-Ruíz, Christopher Bergman, Coralie Boitard, Paolo Boscato, Miguel Caparrós, Nicholas J. Conard, Christelle Draily, Alain Froment, Bertila Galván, Paolo Gambassini, Alejandro Garcia-Moreno, Stefano Grimaldi, Paul Haesaerts, Brigitte Holt, Maria-Jose Iriarte-Chiapusso, Arthur Jelinek, "The timing and spatiotemporal patterning of Neanderthal disappearance", *Nature*j, 512, 306–309 (21 August 2014) doi:10.1038/nature13621 (accessed September 9, 2017).

125. Policarp Hortolàbc, Bienvenido Martínez-Navarroabc, "The Quaternary megafaunal extinction and the fate of Neanderthals: An integrative working hypothesis" Quaternary International, Volume 295, 8 May 2013, Pages 69-72, https://doi.org/10.1016/j.quaint.2012.02.037 (accessed January 2, 2018).

126. Richard G. Klein, "Anatomy, Behavior, and Modern Human Origins", Klein, R.G. Journal of World Prehistory, volume 9, issue 2, 1995, https://doi.org/10.1007/BF02221838 (accessed October 11, 2017), page 167.

K. Boyle, Katherine V. Boyle, Ofer Bar-Yosef, Chris Stringer, Paul Mellars, (editors), Rethinking the Human Revolution: New Behavioural and Biological Perspectives on the Origin and Dispersal of Modern Humans Donald Institute for Archaeological Research, University of Cambridge, England, 2007.

127. Erin Wayman, "When Did the Human Mind Evolve to What It is Today?", smithsonian.com June 25, 2012, https://www.smithsonianmag.com/sci-

ence-nature/when-did-the-human-mind-evolve-to-what-it-is-today-140507905/ (accessed October 11, 2017).

Richard G. Klein, "Anatomy, Behavior, and Modern Human Origins", Klein, R.G. Journal of World Prehistory, volume 9, issue 2, 1995, https://doi.org/10.1007/BF02221838 (accessed October 11, 2017), page 167.

John J. Shea, "Homo sapiens is as Homo sapiens was: Behavioral variability vs. 'behavioral modernity' in Paleolithic archaeology", Current Anthropology, 2011; 52 (1): 1 DOI: 10.1086/658067 (accessed September 3, 2017).

128. Powell A, Shennan S, Thomas MG, "Late Pleistocene demography and the appearance of modern human behavior", Science, 2009 Jun 5;324(5932):1298-301. doi: 10.1126/science.1170165. (accessed October 11, 2017).

129. Mcbrearty, S and Brooks AS, "The revolution that wasn't: a new interpretation of the origin of modern human behavior", Journal of Human Evolution, 2000, Nov;39(5), p. 453.

130. United Nations, "World population projected to reach 9.8 billion in 2050, and 11.2 billion in 2100", June 21, 2017, https://www.un.org/development/desa/en/news/population/world-population-prospects-2017.html (accessed October 18, 2017).

131. López, S., van Dorp, L., & Hellenthal, G., "Human Dispersal Out of Africa: A Lasting Debate," Evolutionary Bioinformatics Online, 2015, 11(Suppl 2), 57–68. http://doi.org/10.4137/EBO.S33489 (accessed November 15, 2017).

Christopher J. Bae, Katerina Douka, Michael D. Petraglia, et al., "On the origin of modern humans: Asian perspectives," Science, 08 Dec 2017: Vol. 358, Issue 6368, eaai9067, DOI: 10.1126/science.aai9067 (accessed January 23, 2018).

132. Jean-Jacques Hublin, and Abdelouahed Ben-Ncer, et al, "New fossils from Jebel Irhoud, Morocco and the pan-African origin of Homo sapiens", Nature, v. 546, June 2017, pp. 289–292, doi:10.1038/nature22336 (accessed November 1, 2017).

133. Athreya S, Wu X., "A multivariate assessment of the Dali hominin cranium from China: Morphological affinities and implications for Pleistocene evolution in East Asia", American Journal of Physical Anthropoly, 2017;00:1–22. https://doi.org/10.1002/ajpa.23305 (accessed November 15, 2017).

134. David Despain, "Early Humans Used Brain Power, Innovation and Teamwork to Dominate the Planet" Scientific American, February 27, 2010, https://www.scientificamerican.com/article/humans-brain-power-origins/ (accessed November 1, 2017).

135. Ethnologue: Languages of the World, https://www.ethnologue.com/ (accessed October 29, 2017).

136. Simon Kemp, "Number of social media users passes 3 billion with no signs of slowing", TNW, https://thenextweb.com/contributors/2017/08/07/number-social-media-

users-passes-3-billion-no-signs-slowing/#.tnw_Oi2CUuNd (accessed November 7, 2017).

137. David Ingram, "Facebook Hits 2 Billion-User Mark, Doubling in Size Since 2012," Reuters, June 27, 2017, https://www.reuters.com/article/us -facebook-users-idUSKBN19I2GG (accessed November 3, 2017)

138. Jonathan Foley, "The Other Inconvenient Truth: The Crisis in Global Land Use", Yale Environment 360, October 5, 2009, http://e360.yale.edu/features/the_other_inconvenient_truth_the_crisis_in_global_land_use (accessed September 19, 2017).

139. Ibid.

140. F. A. Azevedo, L. R. Carvalho, L. T. Grinberg, J. M. Farfel, R. E. Ferretti, R. E. Leite, W. Jacob Filho, R. Lent, and S. Herculano-Houzel, "Equal Numbers of Neuronal and Nonneuronal Cells Make the Human Brain an Isometrically Scaled-Up Primate Brain," Journal of Comparative Neurology 513, no. 5 (April 10, 2009): 532–41, http://www.ncbi.nlm.nih.gov/pubmed?Db=pubmed&Cmd=ShowDetailView&TermToSearch=19226510 (accessed September 19, 201).

141. Donald Brown, *Human Universals*, (New York, NY: McGraw-Hill, 1991).

142. Bloomberg News, "People daydream almost half the day, Harvard study finds", November 14, 2010, http://www.nola.com/health/index.ssf/2010/11/people_daydream_almost_half_th.html (accessed November 15, 2017).

143. Priya Moorjani, Carlos Eduardo G. Amorim, Peter F. Arndt, and Molly Przeworskia, "Variation in the molecular clock of primates", Proceedings of the National Academy of Sciences of the U S A, 2016 Sep 20; 113(38): 10607–10612, doi: 10.1073/pnas.1600374113 (accessed October 29, 2017).

144. United Nations Department of Economic and Social Affairs, "United Nations World Population Prospects: 2015 revision", 29 July 2015, https://esa.un.org/unpd/wpp/publications/Files/WPP2015_Volume-I_Comprehensive-Tables.pdf (accessed November 1, 2017).

145. "Homo sapiens: IUCN Red List of Threatened Species", IUCN, http://www.iucnredlist.org/details/136584/0 (accessed November 1, 2017).

146. Ibid, Klein, Richard G.

147. Tennie C., Call J., Tomasello M., "Ratcheting up the ratchet: on the evolution of cumulative culture", *Philos Trans R Soc Lond B Biol Sci.*, 2009 Aug 27;364(1528):2405-15. doi: 10.1098/rstb.2009.0052 (accessed November 1, 2017).

148. Samuel Bowles and Herbert Gintis, *A Cooperative Species: Human Reciprocity and Its Evolution*, Princeton University Press, 2011, p. 95.

149. Aviroop Biswas, Paul I. Oh, Guy E. Faulkner, Ravi R. Bajaj, Michael A. Silver, Marc S. Mitchell, and David A. Alter, "Sedentary Time and Its Association with Risk for Disease Incidence, Mortality, and Hospitalization in Adults: A Systematic Review and Meta-analysis," Annals of Internal Medicine

(2015), doi:10.7326/M14-1651 (accessed September 14, 2017); E. G. Wilmot, C. L. Edwardson, F. A. Achana, M. J. Davies, T. Gorely, L. J. Gray, K. Khunti, T. Yates, and S. J. H. Biddle, "Sedentary Time in Adults and the Association with Diabetes, Cardiovascular Disease and Death: Systematic Review and Meta-analysis," Diabetologia 55, no. 11 (2012): 2895, doi:10.1007/s00125-012-2677-z (accessed September 14, 2017).

150. Marily Oppezzo, Daniel L. Schwartz, "Give Your Ideas Some Legs: The Positive Effect of Walking on Creative Thinking," *Journal of Experimental Psychology*, 2014, Vol. 40, No. 4, 1142–1152. http://www.apa.org/pubs/journals/releases/xlm-a0036577.pdf (accessed January 20, 2018).

151. Ibid.

152. Dennis Bramble, Daniel Liberman, "Endurance running and the evolution of Homo", *Nature*. 2004 Nov 18;432, (7015):345-52.

153. David R. Carrier, A. K. Kapoor, Tasuku Kimura, Martin K. Nickels, Satwanti, Eugenie C. Scott, Joseph K. So, and Erik Trinkaus, "The Energetic Paradox of Human Running and Hominid Evolution", *Current Anthropology*, 1984 25:4, 483-495

154. Louis Liebenberg, "The relevance of persistence hunting to human evolution", *Journal of human evolution*, Volume 55, Issue 6, December 2008, pp. 1156-1159.

155. Mark Mattson, "Evolutionary aspects of human exercise--born to run purposefully", *Ageing Research Reviews*, 2012 Jul;11(3):347-52. doi: 10.1016/j.arr.2012.01.007. (accessed October 16, 2017).

156. Mark Mattson, "Evolutionary aspects of human exercise--born to run purposefully", *Ageing Research Reviews*, 2012 Jul;11(3):347-52. doi: 10.1016/j.arr.2012.01.007. (accessed October 16, 2017).

157. Paul E. Bendheim, *The Brain Training Revolution: A Proven Workout for Healthy Brain Aging*. Naperville, IL: Sourcebooks, 2009. p. 53.

158. Blaise Williams, "Reduced risk of incident kidney cancer from walking and running", *Medicine & Science in Sports & Exercise,* 2014 Feb; 46(2):312-7. doi: 10.1249/MSS.0b013e3182a4e89c (accessed September 22, 2017).

159. Blaise Williams, "Reduced risk of brain cancer mortality from walking and running", *Medicine & Science in Sports & Exercise*, 2014; 46(5):927-32. doi: 10.1249/MSS.0000000000000176 (accessed September 22, 2017).

160. World Health Organization, "World Heart Day 2017", http://www.who.int/cardiovascular_diseases/world-heart-day-2017/en/ (accessed October 16, 2017).

161. World Health Organization, "Physical Activity", http://www.who.int/topics/physical_activity/en/ (accessed October 16, 2017).

162. Raichlen DA, Pontzer H, Harris JA, et al. Physical activity patterns and biomarkers of cardiovascular disease risk in hunter-gatherers. *Am J Hum Biol*. 2017;29:e22919. https://doi.org/10.1002/ajhb.22919 (accessed October 16, 2017).

163. Herman Pontzer, David A. Raichlen, Brian M. Wood, Audax Z. P. Mabulla, Susan B. Racette, Frank W. Marlowe, "Hunter-Gatherer Energetics and Human Obesity" *PLOS One*, July 25, 2012, https://doi.org/10.1371/journal. pone.0040503 (accessed October 16, 2017).

164. There were exceptions, of course. Some archaeological sites show that small prehistoric populations settled in areas that offered consistent food supply. Some coastal areas, for example, offered sufficient resources for long-term habitation. See: Geoff Bailey, John Parkington, editors, *The Archaeology of Prehistoric Coastlines*, (Cambridge, UK: Cambridge University Press, 1988), p. 9.

165. United Nations, "World's population increasingly urban with more than half living in urban areas", 10 July 2014, http://www.un.org/en/development/desa/news/population/world-urbanization-prospects-2014.html (accessed October 16, 2017).

166. National Genome Research Institute, "Frequently Asked Questions About Genetic and Genomic Science", https://www.genome.gov/19016904/faq-about-genetic-and-genomic-science/ (accessed September 30, 2017)

167. Jeremy Thomson, "Humans did come out of Africa, says DNA", Nature, 7 December 2000, doi:10.1038/news001207-8 http://www.nature.com/news/1998/001207/full/news001207-8.html (accessed January 9, 2018).

168. Ray Bradbury, *Bradbury Speaks: Too Soon from the Cave, Too Far from the Stars*, (New York, NY: William Morrow, 2005), pp. X, XI.

CHAPTER 4

169. Joel Levy, *A Bee in a Cathedral* (Buffalo, NY: Firefly Books, 2011), p. 61.

170. Ibid.

171. David Kestenbaum, " Atomic Tune-Up: How the Body Rejuvenates Itself, NPR, July 14, 2007, https://www.npr.org/templates/story/story.php?storyId=11893583 (accessed January 2, 2018).

172. Ibid.

173. Jamie Trosper, "Why Physics Says You Can Never Actually Touch Anything", Futurism, June 17, 2014 https://futurism.com/why-you-can-never-actually-touch-anything/ (accessed January 16, 2018).

174. Brian Greene, *The Hidden Reality*, (New York, NY: First Vintage Books, 2011) p. 181.

175. New Scientist, *The Quantum World: The Disturbing Reality at the Heart of Reality* (Boston, MA: Nicholas Brealey Publishing, 2017) p. 64.

176. C. J. Davisson, L. H. Germer, "Reflection of Electrons by a Crystal of Nickel", Proceedings of the National Academy of Sciences, U S A. 1928

Apr; 14(4): 317–322, https://www.ncbi.nlm.nih.gov/pmc/articles/PMC1085484/ (accessed January 16, 2018).

177. Avery Thompson, "The Logic-Defying Double-Slit Experiment Is Even Weirder Than You Thought Just by observing the process of the experiment, everything changes", Popular Mechanics, Aug 11, 2016, http://www.popular-mechanics.com/science/a22280/double-slit-experiment-even-weirder/ (accessed January 16, 2018).

178. G. Manning, R. I. Khakimov, R. G. Dall, A. G. Truscott, "Wheeler's delayed-choice gedanken experiment with a single atom", *Nature Physics*, 2015; DOI: 10.1038/nphys3343.

ScienceDaily, "Experiment confirms quantum theory weirdness", 27 May 2015. www.sciencedaily.com/releases/2015/05/150527103110.htm (accessed December 28, 2017).

179. Juan Yin, Yuan Cao, et all "Satellite-based entanglement distribution over 1200 kilometers", *Science,* 16 Jun 2017: Vol. 356, Issue 6343, pp. 1140-1144, DOI: 10.1126/science.aan3211 http://science.sciencemag.org/content/356/6343/1110 (accessed December 28, 2017).

180. Amir D. Aczel, *Entanglement* (New York: Plume ,2003), pp. 249-0.

181. Caleb Scharf, *The Zoomable Universe* (New York, NY: Scientific American, 2017), p. 159.

182. Ibid.

183. Ali Sundermier, "The Particle Physics of You", Symmetry, https://www.symmetrymagazine.org/article/the-particle-physics-of-you11/03/15 (accessed December 29, 2017).

CHAPTER 5

184. Ferris Jabr, "Why Life Does Not Really Exist", *Scientific American*, December 2, 2013, https://blogs.scientificamerican.com/brainwaves/why-life-does-not-really-exist/ (accessed December 29, 2017).

185. Peter Dockrill, "The Largest Study of Life Forms Ever Has Estimated That Earth Is Home to 1 TRILLION Species," Science Alert, 3 May 2016, https://www.sciencealert.com/the-largest-study-of-life-forms-ever-has-estimated-that-earth-is-home-to-1-trillion-species (accessed January 28, 2018).

186. Beverly Peterson Stearns and Stephen C. Stearns, *Watching from the Edge of Extinction*, Yale University Press, 2000, p. x.

Richard Dawkins, *Science in the Soul: Selected Writings of a Passionate Rationalist*, (Bantam Press, London), p. 71.

187. Laetitia Plaisance, M. Julian Caley, Russell E. Brainard, et al., "The Diversity of Coral Reefs: What Are We Missing?" *PLOS One*, October 13, 2011, https://doi.org/10.1371/journal.pone.0025026 (accessed January 27, 2018).

188. UNESCO, "Assessment: World Heritage coral reefs likely to disappear by 2100 unless CO2 emissions drastically reduce," 23 June 2017, http://whc.unesco.org/en/news/1676/ (accessed January 27, 2018).

189. World Wildlife Fund, "What animals live in the Amazon? And 8 other Amazon facts," https://www.worldwildlife.org/stories/what-animals-live-in-the-amazon-and-8-other-amazon-facts (accessed January 27, 2018).

190. Rika Berenguer, Joice Ferreira, Toby Alan Gardner, et al., "A Large-Scale Field Assessment of Carbon Stocks in Human-Modified Tropical Forests," *Global Change Biology*, Volume 20, Issue 12, December 2014, Pages 3713–3726, DOI: 10.1111/gcb.12627 (accessed January 27, 2018).

191. Royal Botanic Gardens, Kew, "State of the World's Plants 2017," https://stateoftheworldsplants.com/ (accessed January 28, 2018).

192. Ibid.

193. Stefan Bengtson, Therese Sallstedt, et al., "Three-dimensional preservation of cellular and subcellular structures suggests 1.6 billion-year-old crown-group red algae," PLOS One, March 14, 2017, https://doi.org/10.1371/journal.pbio.2000735 (accessed January 28, 2018).

194. ScienceDaily, "New Ant Species Discovered In The Amazon Likely Represents Oldest Living Lineage Of Ants," 16 September 2008, www.sciencedaily.com/releases/2008/09/080915174538.htm (accessed January 27, 2018).

195. Encyclopedia Britannica, "Fern," https://www.britannica.com/plant/fern (accessed January 28, 2018).

196. E. Beech, M. Rivers, S. Oldfield, P. P. Smith, "GlobalTreeSearch: The first complete global database of tree species and country distributions," *Journal of Sustainable Forestry*, pp. 454-489, https://doi.org/10.1080/10549811.2017.1310049 (accessed January 28, 2018).

197. Nic Fleming, "Plants Talk to Each Other Using an Internet of Fungus," BBC, November 11, 2014, http://www.bbc.com/earth/story/20141111-plants-have-a-hidden-internet (accessed January 27, 2018).

198. Matthew S. Dodd, Dominic Papineau, et al., "Evidence for early life in Earth's oldest hydrothermal vent precipitates," *Nature*, V. 543, 2 March 2017, doi:10.1038/nature21377 (accessed January 27, 2018).

199. William J. Ripple, Guillaume Chapron, et al., "Saving the World's Terrestrial Megafauna," *BioScience*, Volume 66, Issue 10, 1 October 2016, Pages 807–812, https://doi.org/10.1093/biosci/biw092 (accessed January 28, 2018).

200. Stephen Hart, "Eukaryotic Origins: Revolution in the Classification of Life", *Astrobiology Magazine*, July 31, 2002, https://www.astrobio.net/origin-and-evolution-of-life/eukaryotic-origins-revolution-in-the-classification-of-life/ (accessed December 30, 2017).

Fiona Macdonald, "We Might Finally Have Found Where Complex Life Came From", Science Alert, 13 January 2017, https://www.sciencealert.com/scientists-might-have-finally-found-the-common-ancestor-of-all-complex-life (accessed January 9, 2018).

201. Russell H. Vreeland, William D. Rosenzweig, Dennis W. Powers, "Isolation of a 250 million-year-old halotolerant bacterium from a primary salt crystal," *Nature* 407, 897–900 (19 October 2000) doi:10.1038/35038060 (accessed January 18, 2018).

Jeff Long, "Scientists Rouse Bacterium From 250-million-year Slumber," *Chicago Tribune*, October 19, 2000, http://articles.chicagotribune.com/2000-10-19/news/0010190207_1_bacterium-scientists-solar-system (accessed January 18, 2018).

202. Miriam Kramer, "How Worms Survived NASA's Columbia Shuttle Disaster", Space.com, January 30, 2013, https://www.space.com/19538-columbia-shuttle-disaster-worms-survive.html (accessed December 30, 2017).

203. Rachel Courtland "'Water bears' are first animal to survive space vacuum", *New Scientist*, 8 September 2008, https://www.newscientist.com/article/dn14690-water-bears-are-first-animal-to-survive-space-vacuum/#.U-OzhIBdWd4 (accessed December 30, 2017).

204. NASA, "About Life Detection", https://astrobiology.nasa.gov/research/life-detection/about/ (accessed December 29, 2017).

205. David Enard, Le Cai, et al., "Viruses are a dominant driver of protein adaptation in mammals," eLife, 2016;5:e12469 doi: 10.7554/eLife.12469 https://elifesciences.org/articles/12469 (accessed January 18, 2018).

206. Belshaw R, Pereira V, Katzourakis A, Talbot G, Paces J, Burt A, Tristem M (April 2004). "Long-term reinfection of the human genome by endogenous retroviruses". *Proceedings of the National Academy of Sciences*. 101 (14): 4894–9. doi:10.1073/pnas.0307800101 (accessed November 3, 2017).

207. Arshan Nasir, Kyung Mo Kim, Gustavo Caetano-Anolles, "Giant viruses coexisted with the cellular ancestors and represent a distinct supergroup along with superkingdoms Archaea, Bacteria and Eukarya", *BMC Evolutionary Biology*, 2012; 12 (1): 156 DOI: 10.1186/1471-2148-12-156

Arshan Nasir and Gustavo Caetano-Anollés, "A phylogenomic data-driven exploration of viral origins and evolution", Science Advances, September 2015 DOI: 10.1126/sciadv.1500527

University of Illinois at Urbana-Champaign, "Study adds to evidence that viruses are alive." ScienceDaily, ScienceDaily, 25 September 2015, www.sciencedaily.com/releases/2015/09/150925142658.htm (accessed December 29, 2017).

208. Ibid, Jabr.

209. Takayuki Tashiro, Akizumi Ishida, et al, "Early trace of life from 3.95 Ga sedimentary rocks in Labrador, Canada", Nature, 549, 516–518, 28 September 2017, doi:10.1038/nature24019 (accessed December 29, 2017).

210. Peter Ward, and Joe Kirschvink, A New History of Life: The Radical New Discoveries about the Origins and Evolution of Life on Earth (New York, NY: Bloomsbury Press, 2015), p. 31.

211. Science Daily, "Early trilobites had stomachs, new fossil study finds:

Remarkable Chinese specimens contradict previous assumptions about trilo-bite digestive systems and evolution." ScienceDaily, 21 September 2017, www.sciencedaily.com/releases/2017/09/170921141201.htm (accessed January 5, 2018).

212. S. Xiao, X. Yuan, A. H. Knoll, "Eumetazoan fossils in terminal Pro-terozoic phosphorites?" Proceedings of the National Academy of Sciences of the United States of America, 97(25), 2000, 13684–13689. https://www.ncbi.nlm.nih.gov/pmc/articles/PMC17636/ (accessed January 5, 2018).

213. Douglas Fox, "What sparked the Cambrian explosion?", Nature 530, pp. 268–270, (18 February 2016), doi:10.1038/530268a https://www.nature.com/news/what-sparked-the-cambrian-explosion-1.19379 (accessed January 5, 2018).

214. Ibid, Ward, Kirschvink.

215. Jeremie Palacci, Stefano Sacanna, et al., "Living Crystals of Light-Activated Colloidal Surfers," Science, 22 Feb 2013: Vol. 339, Issue 6122, pp. 936-940, DOI: 10.1126/science.1230020 (accessed January 18, 2018).

216. Ibid, Jabr.

217. Brian Koberlein, interview with the author, December 15, 2017.

218. Theodosius Dobzhansky, "Nothing in Biology Makes Sense except in the Light of Evolution", The American Biology Teacher, Vol. 35 No. 3, Mar., 1973; (pp. 125-129) DOI: 10.2307/4444260 (accessed December 30, 2017).

219. Cameron M. Smith, The Fact of Evolution (Amherst, NY: Prometheus Books, 2011), p. 22.

220. Art Swift, "In U.S., Belief in Creationist View of Humans at New Low", Gallup, May 22, 2017, http://news.gallup.com/poll/210956/belief-cre-ationist-view-humans-new-low.aspx (accessed January 5, 2018).

221. Guy P. Harrison, "The Dinosaur Hunter", Caymanian Compass, Sep-tember 14, 2001, p. B6.

222. Guy P. Harrison, "Lucy in the Sky", Caymanian Compass, August 23, 2002, p. A11.

223. Guy P. Harrison, "Who's Your Daddy?", Caymanian Compass, October 17, 2003, p. A1.

224. Ibid, Jabr.

225. Ann M O'Hara, Fergus Shanahan. "The Gut Flora as a Forgotten Organ." EMBO Reports, 7.7, 2006, pp. 688–693, https://www.ncbi.nlm.nih.gov/pmc/articles/PMC1500832/ (accessed January 24, 2018).

226. Ibid.

227. Gaorui Bian, Gregory B. Gloor, Aihua Gong, et al., "The Gut Micro-biota of Healthy Aged Chinese Is Similar to That of the Healthy Young," mSphere, 2017; 2 (5): e00327-17 DOI: 10.1128/mSphere.00327-17 (accessed January 26, 2018).

228. ScienceDaily, "'Ridiculously healthy' elderly have the same gut microbiome as healthy 30-year-olds." 11 October 2017. www.sciencedaily.com/

releases/2017/10/171011123728.htm (accessed January 26, 2018).

229. Jason Lloyd-Price, Galeb Abu-Ali, Curtis Huttenhower, "The healthy human microbiome," Genome Medicine, 20168:51, 27 April 2016, doi. org/10.1186/s13073-016-0307-y (accessed January 26, 2018).

230. L. Galland, "The Gut Microbiome and the Brain," Journal of Medicinal Food. 2014;17(12):1261-1272. doi:10.1089/jmf.2014.7000. (accessed January 24, 2018).

S. Devaraj, P. Hemarajata, J. Versalovic, "The Human Gut Microbiome and Body Metabolism: Implications for Obesity and Diabetes," Clinical chemistry, 2013;59(4):617-628. doi:10.1373/clinchem.2012.187617. (accessed January 24, 2018).

231. Austin T. Mudd, Kirsten Berding, Mei Wang, Sharon M. Donovan, Ryan N. Dilger, "Serum cortisol mediates the relationship between fecal Ruminococcus and brain N-acetylaspartate in the young pig," Gut Microbes, 2017; 1 DOI: 10.1080/19490976.2017.1353849 (accessed January 24, 2018).

232. K. Tillisch K, J. Labus J, et al., "Consumption of Fermented Milk Product With Probiotic Modulates Brain Activity," Gastroenterology, 2013;144(7):10.1053/j.gastro.2013.02.043, doi:10.1053/j.gastro.2013.02.043. (accessed January 24, 2018).

David Kohn, "When Gut Bacteria Change Brain Function," The Atlantic, Jun 24, 2015, https://www.theatlantic.com/health/archive/2015/06/gut-bacteria-on-the-brain/395918/ (accessed January 25, 2018).

233. Eamonn M. M. Quigley, "Gut Bacteria in Health and Disease," Gastroenterology & Hepatology, 9(9), 2013, pp. 560–569. https://www.ncbi.nlm.nih.gov/pmc/articles/PMC3983973/ (accessed January 24, 2018).

234. M. J. Hill, "Intestinal flora and endogenous vitamin synthesis," Eur J Cancer Prev, 1997 Mar; 6 Suppl 1:S43-5, https://www.ncbi.nlm.nih.gov/pubmed/9167138 (accessed January 24, 2018).

235. A. Pärtty, M. Kalliomäki, P. Wacklin, et al., "A possible link between early probiotic intervention and the risk of neuropsychiatric disorders later in childhood: a randomized trial," Pediatr Res. 77(6):823-8, 2015 https://www.ncbi.nlm.nih.gov/pubmed/25760553 (accessed January 25, 2018).

236. L. Galland, ibid.

Ed Young, "Gut reaction: the surprising power of microbes," The Guardian, August 25, 2016, https://www.theguardian.com/science/2016/aug/25/gut-reaction-surprising-power-of-microbes (accessed January 26, 2018).

237. Bravo JA, Forsythe P, Chew MV, Escaravage E, Savignac HM, Dinan TG, Bienenstock J, Cryan JF., "Ingestion of Lactobacillus strain regulates emotional behavior and central GABA receptor expression in a mouse via the vagus nerve," Proc Natl Acad Sci U S A. 2011 Sep 20;108(38):16050-5. doi: 10.1073/pnas.1102999108. (accessed January 27, 2018).

Nicola Jones, "Friendly bacteria cheer up anxious mice," Nature, 30 August 2011, https://www.nature.com/news/2011/110830/full/news.2011.510.html#B1

(accessed January 27, 2018).

Ira Flatlow, "Probiotic Bacteria Chill Out Anxious Mice," NPR: Science Friday, September 2, 2011 https://www.npr.org/2011/09/02/140146780/probiotic-bacteria-chill-out-anxious-mice (accessed January 27, 2018).

238. Rasnik K. Singh, Hsin-Wen Chang, et al., "Influence of diet on the gut microbiome and implications for human health," Journal of translational medicine," April 8, 2017; 15: 73, doi: 10.1186/s12967-017-1175-y (accessed January 17, 2018).

239. S. Gupta, E. Allen-Vercoe, E. Petrof, "Fecal microbiota transplantation: in perspective," Therapeutic Advances in Gastroenterology, 2016;9(2):229-239. doi:10.1177/1756283X15607414. (accessed January 26, 2018).

240. Mike Orcutt, "Companies Aim to Make Drugs from Bacteria That Live in the Gut," MIT Technology Review, January 18, 2016, https://www.technologyreview.com/s/545446/companies-aim-to-make-drugs-from-bacteria-that-live-in-the-gut/ (accessed January 26, 2018).

241. Emily Mullin, "Gut Check: Scientists are Wary of At-Home Microbiome Tests," MIT Technology Review, March 24, 2017, https://www.technologyreview.com/s/603900/gut-check-scientists-are-wary-of-at-home-microbiome-tests/ (accessed January 26, 2018).

242. Begoña Cerdá, et al., "Gut Microbiota Modification: Another Piece in the Puzzle of the Benefits of Physical Exercise in Health?" Front Physiol. 2016; 7: 51, 2016 Feb 18. doi: 10.3389/fphys.2016.00051 (accessed January 17, 2018).

Vincenzo Monda, Ines Villano, Antonietta Messina, et al., "Exercise Modifies the Gut Microbiota with Positive Health Effects," *Oxidative Medicine and Cellular Longevity*, vol. 2017, Article ID 3831972, 2017. doi:10.1155/2017/3831972 (accessed January 26, 2018).

243. Global Market Insights Inc., "Probiotics Market Size to Exceed USD 64 Billion by 2023," May 10, 2016, https://www.prnewswire.com/news-releases/probiotics-market-size-to-exceed-usd-64-billion-by-2023-global-market-insights-inc-578769201.html (accessed January 17, 2018).

244. Rasnik K. Singh, Hsin-Wen Chang, et al., "Influence of diet on the gut microbiome and implications for human health," Journal of translational medicine," April 8, 2017; 15: 73, doi: 10.1186/s12967-017-1175-y (accessed January 17, 2018).

245. Angela Marcobal, Mark Underwood, David Mills, "Rapid Determination of the Bacterial Composition of Commercial Probiotic Products by Terminal Restriction Fragment Length Polymorphism Analysis," Journal of Pediatric Gastroenterology and Nutrition, May 2008 - Volume 46 - Issue 5 - p 608–611 doi: 10.1097/MPG.0b013e3181660694 (accessed January 26, 2018).

Michael Pollan, "Some of My Best Friends Are Germs," *New York Times*, May 15, 2013, http://www.nytimes.com/2013/05/19/magazine/say-hello-to-the-100-trillion-bacteria-that-make-up-your-microbiome.html?pagewanted=1 (accessed January 26, 2018).

246. John Cryan, interview with the author, January 27, 2018.

247. National Institutes of Health, "Probiotics: In Depth", https://nccih. nih.gov/health/probiotics/introduction.htm (accessed January 25, 2018).

248. Ferranti, Erin et al. "20 Things You Didn't Know About the Human Gut Microbiome." The Journal of cardiovascular nursing 29.6 (2014): 479–481. PMC. https://www.ncbi.nlm.nih.gov/pmc/articles/PMC4191858/#R2 (accessed January 24, 2018).

249. Jeff Leach, "Gut microbiota: Please pass the microbes," Nature, 04 December 2013, http://www.nature.com/articles/504033c (accessed January 26, 2018).

250. Filocamo, C. Nueno-Palop, et al., "Effect of garlic powder on the growth of commensal bacteria from the gastrointestinal tract," Phytomedicine, 2012 Jun 15;19(8-9):707-11. doi: 10.1016/j.phymed.2012.02.018. (accessed January 26, 2018).

251. K. Magnusson, Hauck, et al. "Relationships between diet-related changes in the gut microbiome and cognitive flexibility," Neuroscience, 2015 Aug 6;300:128-40. doi: 10.1016/j.neuroscience.2015.05.016. (accessed January 27, 2018).

252. Lucy Shewell, "Everything you always wanted to know about fermented foods," Science-Based Medicine, October 30, 2015, https://sciencebased-medicine.org/everything-you-always-wanted-to-know-about-fermented-foods/ (accessed January 26, 2018).

253. Rasnik K. Singh, Hsin-Wen Chang, et al., "Influence of diet on the gut microbiome and implications for human health," Journal of translational medicine," April 8, 2017; 15: 73, doi: 10.1186/s12967-017-1175-y (accessed January 17, 2018).

254. Ibid.

255. University of Michigan Health System, "High Fiber Diet," Michigan Bowel Control Program, http://www.med.umich.edu/1libr/MBCP/HighFiber-Diet.pdf (accessed January 26, 2018).

256. Michael Pollan, ibid.

257. Centers for Disease Control, "CDC: 1 in 3 antibiotic prescriptions unnecessary," May 3, 2016, https://www.cdc.gov/media/releases/2016/p0503-unnecessary-prescriptions.html (accessed January 17, 2018).

258. Martin Blaser, "Antibiotic overuse: Stop the killing of beneficial bacteria," Nature, 476, 393–394, 25 August 2011, doi:10.1038/476393a (accessed January 24, 2018).

259. Michael Tennesen, "The Trillions of Microbes That Call Us Home—and Help Keep Us Healthy," Discover, March 2011, http://discovermagazine. com/2011/mar/04-trillions-microbes-call-us-home-help-keep-healthy (accessed January 17, 2018).

260. Les Dethlefsen, David A. Relman, "Incomplete recovery and individualized responses of the human distal gut microbiota to repeated antibiotic per-

turbation," Proceedings of the National Academy of Sciences, September 2010, 201000087; DOI: 10.1073/pnas.1000087107 (accessed January 26, 2018).

261. Camilo Mora, Derek P. Tittensor, Sina Adl, Alastair G. B. Simpson, Boris Worm, "How Many Species Are There on Earth and in the Ocean?" PLOS Biology, August 23, 2011. https://doi.org/10.1371/journal.pbio.1001127 (accessed November 9, 2017).

262. Jack A. Gilbert, Josh D. Neufeld, "Life in a World without Microbes", PLoS Biology, 2014 Dec; 12(12), doi: 10.1371/journal.pbio.1002020 (accessed October 29, 2017).

263. Kenneth J. Locey, Jay T. Lennona, "Scaling laws predict global microbial diversity", Proceedings of the National Academy of Sciences, 5970–5975, vol. 113 no. 21, May 24, 2016, doi: 10.1073/pnas.1521291113 (accessed September 27, 2017).

264. New Scientist, "Thousands of new lifeforms discovered that redraw tree of life", 11 September 2017, https://www.newscientist.com/article/2146942-thousands-of-new-lifeforms-discovered-that-redraw-tree-of-life/

265. Mark Kowarsky, Joan Camunas-Soler, Michael Kertesz, Iwijn De Vlaminck, Winston Koh, Wenying Pan, Lance Martin, Norma F. Neff, Jennifer Okamoto, Ronald J. Wong, Sandhya Kharbanda, Yasser El-Sayed, Yair Blumenfeld, David K. Stevenson, Gary M. Shaw, Nathan D. Wolfe, Stephen R. Quake. "Numerous uncharacterized and highly divergent microbes which colonize humans are revealed by circulating cell-free DNA", Proceedings of the National Academy of Sciences, 2017; 201707009 DOI: 10.1073/pnas.1707009114 (accessed September 16, 2017).

266. James F. Meadow, et al. "Humans Differ in Their Personal Microbial Cloud." Ed. Valeria Souza. PeerJ 3 (2015): e1258. PMC. https://www.ncbi.nlm.nih.gov/pmc/articles/PMC4582947/ (accessed December 30, 2017).

PhysOrg, "New research finds that people emit their own personal microbial cloud", September 22, 2015, https://phys.org/news/2015-09-people-emit-personal-microbial-cloud.html (accessed December 30, 2017).

267. DL Theobald, "A formal test of the theory of universal common ancestry." Nature, 2010 May 13; 465(7295):219-22, doi: 10.1038/nature09014 (accessed December 30, 2017).

268. Lynn Margulis, Dorion Sagan, What is Life? (Berkeley and Los Angeles, CA: University of California, 2000) p. 217.

269. Ibid, p. 216.

CHAPTER 6

270. "Australopithecus aferensis,"Smithsonian National Museum of Natural History, http://humanorigins.si.edu/evidence/human-fossils/species/

australopithecus-afarensis (accessed October 29, 2017).

271. Jean-Jacques Hublin, Abdelouahed Ben-Ncer, et al., "New fossils from Jebel Irhoud, Morocco and the pan-African origin of Homo sapiens,"Nature, 546, 289–292, 08 June 2017, doi:10.1038/nature22336 (accessed January 9, 2018).

272. Wrangham, R., Catching Fire: How Cooking Made Us Human (New York, Basic Books, 2009).

Ferraro, J. V., et al., "Earliest archaeological evidence of persistent hominin carnivory,"PLoS ONE 8(4):e62174, 2013, doi:10.1371/journal.pone.0062174 (accessed January 9, 2018).

Briana Pobiner, "Meat-Eating Among the Earliest Humans,"American Scientist, https://www.americanscientist.org/article/meat-eating-among-the-earliest-humans (accessed January 9, 2018).

273. Bob Holmes, "The ascent of medallion man,"New Scientist, 9 May 1998, https://www.newscientist.com/article/mg15821332-900-the-ascent-of-medallion-man/ (accessed January 9, 2018).

274. Marek Kohn, Steven Mithen, "Handaxes: products of sexual selection?,"Antiquity, Volume 73, Issue 281, September 1999 , pp. 518-526, https://doi.org/10.1017/S0003598X00065078 (accessed January 9, 2018).

275. López, S., van Dorp, L., & Hellenthal, G., "Human Dispersal Out of Africa: A Lasting Debate," Evolutionary Bioinformatics Online, 2015, 11(Suppl 2), 57–68. http://doi.org/10.4137/EBO.S33489 (accessed November 15, 2017).

276. Chao Liu, Yuchun Tang , et al. "Increasing breadth of the frontal lobe but decreasing height of the human brain between two Chinese samples from a Neolithic site and from living humans,"American Journal of Physical Anthropology, Volume 154, Issue 1, May 2014, pp. 94–103.

277. James C. Scott, Against the Grain: A Deep History of the Earliest States, (New Haven, CT: Yale University Press, 2017), p. 69.

278. Mark Lipson, Anna Szécsényi-Nagy, Swapan Mallick, et al, "Parallel palaeogenomic transects reveal complex genetic history of early European farmers,"Nature, 2017; DOI: 10.1038/nature24476 (accessed November 9, 2017).

279. Denis J. Murphy, People, Plants and Genes: The Story of Crops and Humanity, (Oxford, UK: Oxford University Press, 2007), p. 126.

280. Spencer Wells, Pandora's Seed: The Unforeseen Cost of Civilization, (New York, NY: Random House, 2010), p. 24.

281. Ibid, Wells, p. 23-24.

282. Ibid, Wells, p. 22-23.

283. Jared Diamond, Guns, Germs, and Steel: The Fates of Human Societies, (New York, W. W. Norton & Company, 2005), pp. 196-205.

284. International Livestock Research Institute, "Mapping of Poverty and Likely Zoonoses Hotspots, July 2, 2012, https://cgspace.cgiar.org/bitstream/handle/10568/21161/ZooMap_July2012_final.pdf?sequence=4 (accessed

November 17, 2017), p. 12.

285. Ibid.

286. Grace, D., Jones B., McKeever, D., Pfeiffer, D., et al., "Zoonoses: Wildlife/livestock interactions: A report to the Department for International Development, UK,"Submitted by: The International Livestock Research Institute, Nairobi & Royal Veterinary College, London. 2011. Cited in: International Livestock Research Institute, "Mapping of Poverty and Likely Zoonoses Hotspots, July 2, 2012, p. 12.

287. Jared Diamond, "The Worst Mistake in the History of the Human Race,"Discover, May 01, 1999

http://discovermagazine.com/1987/may/02-the-worst-mistake-in-the-history-of-the-human-race (accessed November 15, 2017).

288. Diamond, Ibid.

289. James C. Scott, Against the Grain: A Deep History of the Earliest States, (New Haven, CT: Yale University Press, 2017), pp. 9-10.

290. Ibid, Wells, pp. 118-120.

291. Ibid, pp. 112-113.

292. Vigo, Daniel et al., "Estimating the true global burden of mental illness,"The Lancet Psychiatry, Volume 3, Issue 2, 171 - 178, DOI: http://dx.doi.org/10.1016/S2215-0366(15)00505-2 (accessed January 10, 2018).

293. Brandon H Hidaka, "Depression as a disease of modernity: explanations for increasing prevalence,"Journal of Affect Disord, 2012 Nov; 140(3): 205–214, doi: 10.1016/j.jad.2011.12.036 (accessed January 10, 2018).

294. W. Scheidel. The Great Leveler: Violence and the History of Inequality from the Stone Age to the Twenty-First Century (Princeton Univ. Press, 2017).

295. Timothy A. Kohler, Michael E. Smith, Amy Bogaard, Gary M. Feinman, Christian E. Peterson, Alleen Betzenhauser, Matthew Pailes, Elizabeth C. Stone, Anna Marie Prentiss, Timothy J. Dennehy, Laura J. Ellyson, Linda M. Nicholas, Ronald K. Faulseit, Amy Styring, Jade Whitlam, Mattia Fochesato, Thomas A. Foor, Samuel Bowles, "Greater post-Neolithic wealth disparities in Eurasia than in North America and Mesoamerica,"Nature, 2017; DOI: 10.1038/nature24646 (accessed November 16, 2017).

296. Oxfam, "Just 8 men own same wealth as half the world,"16 January 2017, https://www.oxfam.org/en/pressroom/pressreleases/2017-01-16/just-8-men-own-same-wealth-half-world / (accessed November 16, 2017).

297. Raj Chetty, Michael Stepner, et al, "The Association Between Income and Life Expectancy in the United States,"2001-2014, JAMA, 2016;315(16):1750-1766. doi:10.1001/jama.2016.4226 (accessed November 16, 2017).

298. Walter Scheidel, interview with the author, December 11, 2017.

299. Ibid.

300. Seyfarth, Robert M., Cheney, Dorothy L., Marler, Peter "Vervet monkey alarm calls: Semantic communication in a free-ranging primate". Animal Behaviour, November, 1980, 28 (4): pp. 1070–1094. doi:10.1016/S0003-

3472(80)80097-2. (accessed October 21, 2017).

301. Bolhuis JJ, Tattersall I, Chomsky N, Berwick RC, "How Could Language Have Evolved?,"PLOS Biology, August 26, 2014, 12(8), DOI: 10.1371/journal.pbio.1001934 (accessed November 8, 2017).

302. Anne Trafton, "Neuroscientists identify key role of language gene,"MIT News, September 15, 2014, http://news.mit.edu/2014/language-gene-0915 (accessed November 8, 2017).

303. Ray Jackendoff, "How Did Language Begin?,"Linguistics Society of America, https://www.linguisticsociety.org/resource/faq-how-did-language-begin (accessed November 8, 2017).

304. T. J. H. Morgan, N. T. Uomini, et al, "Experimental evidence for the co-evolution of hominin tool-making teaching and language,"Nature Communications 6, Article number: 6029 (2015), doi:10.1038/ncomms7029 (accessed October 1, 2017).

305. I was persuaded to give the intriguing idea of complex hand/sign language predating oral language more consideration after a conversation with Josephine DeVera. Read more about it here: William C. Stokoe, "Gesture to Language to Speech," The DANA Foundation, October 01, 2001, http://www.dana.org/Cerebrum/Default.aspx?id=39293 (accessed January 19, 2018).

306. Carotenuto F., Tsikaridze N., et al, "Venturing out safely: The biogeography of Homo erectus dispersal out of Africa,"Journal of Human Evolution, 2016 Jun; 95:1-12. doi: 10.1016/j.jhevol.2016.02.005. (accessed November 16, 2017).

307. Daniel L. Everett, How Language Began: The Story of Humanity's Greatest Invention, (New York, NY: Liveright, 2017), p. 15.

308. Everett, Ibid, p. 64.

309. Pinker, S. & Bloom, P., "Natural language and natural selection,"Behavioral and Brain Sciences, 13 (4), 1990: pp. 707-784.

310. Pinker, Ibid.

311. Colin Barras, "Just how are we related to our chimp cousins?,"New Scientist, 16 March 2016, https://www.newscientist.com/article/2081012-just-how-are-we-related-to-our-chimp-cousins/ (accessed September 29, 2017).

Arnason U1, Gullberg A, Janke A., "Molecular timing of primate divergences as estimated by two nonprimate calibration points,"Journal of Molecular Evolution, J Mol Evol. 1998 Dec;47(6):718-27.

https://www.ncbi.nlm.nih.gov/pubmed/9847414 (accessed September 29, 2017).

312. Mark Pagel, "How humans evolved language, and who said what first,"New Scientist, February 3, 2016, https://www.newscientist.com/article/2075666-how-humans-evolved-language-and-who-said-what-first/ (accessed November 1, 2017).

313. Bruce Bower, "Engraved pigments point to ancient symbolic tradition,"Science News, June 12, 2009

https://www.sciencenews.org/article/engraved-pigments-point-ancient-symbolic-tradition (accessed November 27, 2017).

314. Takayuki Tashiro, Akizumi Ishida, Masako Hori, Motoko Igisu, Mizuho Koike, Pauline Méjean, Naoto Takahata, Yuji Sano, Tsuyoshi Komiya, "Early trace of life from 3.95 Ga sedimentary rocks in Labrador, Canada,"Nature, 549, 516–518, 28 September 2017, doi:10.1038/nature24019 (accessed November 18, 2017).

CHAPTER 7

316. National Genome Research Institute, "Frequently Asked Questions About Genetic and Genomic Science", https://www.genome.gov/19016904/faq-about-genetic-and-genomic-science/ (accessed September 30, 2017)

316. American Association of Physical Anthropologists, "AAPA Statement on Biological Aspects of Race", American Journal of Physical Anthropology, vol. 101, pp 569-570, 1996, http://physanth.org/about/position-statements/biological-aspects-race/ (accessed November 22, 2017).

317. Agustin Fuentes, The Creative Spark: How Imagination Made Humans Exceptional (New York, NY: Dutton, 2017), p. 47.

318. Sheldon Krimsky and Kathleen Sloan (editors), Race and the Genetic Revolution: Science, Myth, and Culture, (New York, Columbia University Press 2011), p. 16.

319. Guy P. Harrison, Race and Reality: What Everyone Should Know About Our Biological Diversity. Amherst, NY: Prometheus Books, 2010.

320. Cavalli-Sforza, Luigi Luca; Menozzi, Paolo; Piazza, Alberto. The History and Geography of Human Genes, (Princeton, NJ: Princeton University Press, 1994), p. 19.

321. Jennifer K. Wagner, et al. "Anthropologists' Views on Race, Ancestry, and Genetics." American Journal of Physical Anthropology, 162(2), February 2017, 318–327. http://doi.org/10.1002/ajpa.23120 (accessed November 22, 2017).

322. American Anthropology Association, "AAA Statement on Race", adopted May 17, 1998, http://www.americananthro.org/ConnectWithAAA/Content.aspx?ItemNumber=2583 (accessed November 22, 2017)

323. Jacqueline M. Chen, David L. Hamilton, "Natural ambiguities: Racial categorization of multiracial individuals", Journal of Experimental Social Psychology, August 2013, http://www.apa.org/news/press/releases/2013/08/racial-categorization.pdf (accessed November 22, 2017).

324. Steve Bradt, "One-drop rule' persists", Harvard Gazette, December 9, 2010, https://news.harvard.edu/gazette/story/2010/12/one-drop-rule-persists/ (accessed November 22, 2017).

325. Ibid, xxLuigi Luca Cavalli-Sforza; Paolo Menozzi; Alberto Piazza, *The History and Geography of Human Genes* (Princeton, NJ: Princeton University Press, 1994, p. 19.

326. Ibid, American Association of Physical Anthropologists.

327. Stephan C. Schuster, Webb Miller, Vanessa M. Hayes, et al, "Complete Khoisan and Bantu genomes from southern Africa", Nature, 463, 943–947, 18 February 2010, doi:10.1038/nature08795 (accessed November 23, 2017).

328. Charles N Rotimi, "Are medical and nonmedical uses of large-scale genomic markers conflating genetics and 'race'?", Nature Genetics, 36, S43–S47, 2004, doi:10.1038/ng1439 http://www.nature.com/articles/ng1439 (accessed November 23, 2017).

329. Nicholas Wade, A Troublesome Inheritance: Genes, Race, and Human History (New York, NY: Penguin, 2014), p. 121.

330. Stanford Center for Computational, Evolutionary and Human Genomics, "Letters: 'A Troublesome Inheritance'", https://cehg.stanford.edu/letter-from-population-geneticists (accessed November 26, 2017).

331. Alan Templeton, "Human Races: A Genetic and Evolutionary Perspective," American Anthropologist, Volume 100, Number 3, 1 September 1998, pp. 632-650, DOI: https://doi.org/10.1525/aa.1998.100.3.632 (accessed January 26, 2018).

332. Michael Yudell, Dorothy Roberts, Rob DeSalle, Sarah Tishkoff, "Taking race out of human genetics", Science 05 Feb 2016: Vol. 351, Issue 6273, pp. 564-565 DOI: 10.1126/science.aac4951 (accessed January 27, 2018).

333. Ibid.

334. Ibid, Rotimi.

335. Ibid, Rotimi.

336. PBS, "Race: The Power of an Illusion", http://www.pbs.org/race/000_About/002_04-experts-01-10.htm (accessed November 24, 2017).

337. Vence L. Bonham, Shawneequa L. Callier, Charmaine D. Royal, "Will Precision Medicine Move Us beyond Race?," New England Journal of Medicine, 2016; 374 (21): 2003 DOI: 10.1056/NEJMp1511294 (accessed January 17, 2018).

Michael Bigby, Diane Thaler, "Describing patients' 'race' in clinical presentations should be abandoned", *Journal of the American Academy of Dermatology*, June 2006, Volume 54, Issue 6, Pages 1074–1076, DOI: http://dx.doi.org/10.1016/j.jaad.2005.10.067 (accessed January 17, 2018).

Roger Collier, "A race-based detour to personalized medicine," CMAJ, 2012 Apr 17; 184(7): E351–E353. doi: 10.1503/cmaj.109-4133 (accessed January 17, 2018).

338. Troy Duster, "Race and Reification in Science," Science, 18 Feb 2005: Vol. 307, Issue 5712, pp. 1050-1051, DOI: 10.1126/science.1110303 (accessed January 17, 2018).

Guy P. Harrison, *Race and Reality: What Everyone Should Know About*

Our Biological Diversity (Amherst, NY: Prometheus Books, 2010), pp. 161-186.

339. It is important to note that cultural race groups do exist and they do impact the health of people. Being a member of an abused or neglected cultural group of any kind can have severe negative health consequences. For more on this see the chapter, "Are You a Patient or a Color" in my book, Race and Reality: What Everyone Should Know About Our Biological Diversity, pp. 161-186.

340. Ritchie Witzig, "The Medicalization of Race: Scientific Legitimization of a Flawed Social Construct," Annals of Internal Medicine 125, no. 8 (October 15, 1996): 675–79, www.annals.org/cgi/content/full/125/8/675 (accessed January 17, 2018).

341. Jewish News of Northern California, "Now more common in non-Jews: Years of genetic testing virtually erase Tay-Sachs", June 13, 2003, https://www.jweekly.com/2003/06/13/now-more-common-in-non-jews-years-of-genetic-testing-virtually-erase-tay-sa/ (accessed November 26, 2017).

342. Buckles, Julie "The Success Story of Gene Tests", *Genome News Network*, August 20, 2001, http://www.genomenewsnetwork.org/articles/08_01/Tay_Sachs_gene_tests.shtml (accessed November 26, 2017).

343. T. E. Kelly, G. A. Chase, et al, "Tay-Sachs disease: high gene frequency in a non-Jewish population", American Journal of Human Genetics, 1975 May; 27(3): 287–291. (accessed September 9, 2017).

344. Wikipedia, "100 Metres at the World Championships in Athletics". https://en.wikipedia.org/wiki/100_metres_at_the_World_Championships_in_Athletics (accessed November 29, 2017).

345. Doug Merlino, "Magic Johnson and Larry Bird: The Rivalry That Transformed the NBA", Bleacher Report, May 13, 2011, http://bleacherreport.com/articles/699088-magic-johnson-and-larry-bird-the-rivalry-that-transformed-the-nba (accessed November 27, 2017).

346. Harrison, Race and Reality, pp. 113-159; and Cynthia M. Frisby, How You See Me, How You Don't, (Mustang, OK: Tate, 2015), pp. 385-409.

CHAPTER 8

347. Rhett Herman, Scientific American, "How Fast is the Earth Moving?," https://www.scientificamerican.com/article/how-fast-is-the-earth-mov/ (accessed December 1, 2017).

348. Greg Simpson, interview with the author, January 29, 2018.

349. International Astronomical Union, "Pluto and the Developing Landscape of Our Solar System" https://www.iau.org/public/themes/pluto/ (accessed December 23, 2017).

350. Neil DeGrasse Tyson, Welcome to the Universe: An Astrophysical

Tour, (Princeton, NJ: Princeton University Press, 2016), p. 142.

351. International Astronomical Union, "RESOLUTION B5 Definition of a Planet in the Solar System" https://www.iau.org/static/resolutions/Resolution_GA26-5-6.pdf (accessed December 23, 2017).

352. Paul Rincon "Why is Pluto no longer a planet?" BBC http://www.bbc.com/news/science-environment-33462184 (accessed December 23, 2017).

Philip Metzger, "Nine Reasons Why Pluto Is a Planet" April 13, 2015 http://www.philipmetzger.com/blog/nine-reasons-why-pluto-is-a-planet/ (accessed December 23, 2017).

353. NASA, "What Is Jupiter?" July 8, 2016 https://www.nasa.gov/audience/forstudents/k-4/stories/nasa-knows/what-is-jupiter-k4.html (accessed December 23, 2017).

354. Ibid.

355. Konstantin Batygin, Michael E. Brown, "Evidence for a Distant Giant Planet in the Solar System", The Astronomical Journal, Volume 151, Number 2, January 20, 2016 http://iopscience.iop.org/article/10.3847/0004-6256/151/2/22 (accessed December 23, 2017).

NASA, "Hypothetical 'Planet X': In Depth" https://solarsystem.nasa.gov/planets/planetx/indepth (accessed December 23, 2017).

356. Ibid.

357. By Richard Hollingham, "The Search for the Solar System's Most Likely Places for Life" BBC Future, 17 July 2017, http://www.bbc.com/future/story/20170717-the-search-for-the-solar-systems-second-genesis (accessed January 12, 2018).

358. NASA, "NASA Missions Provide New Insights into 'Ocean Worlds' in Our Solar System" April 13, 2017 https://www.nasa.gov/press-release/nasa-missions-provide-new-insights-into-ocean-worlds-in-our-solar-system (accessed December 23, 2017).

359. NASA, "Enceladus: Ocean Moon" https://saturn.jpl.nasa.gov/science/enceladus/ (accessed December 23, 2017).

360. NASA, "Europa: 10 Need-To-Know Things" https://solarsystem.nasa.gov/planets/europa/needtoknow (accessed December 23, 2017).

361. For a list of confirmed and prospective moons, see: NASA, "Our Solar System: Moons" https://solarsystem.nasa.gov/planets/solarsystem/moons (accessed December 23, 2017).

362. NASA, "Asteroids in Depth" https://solarsystem.nasa.gov/planets/asteroids/indepth (accessed December 23, 2017).

363. Goddard Space Flight Center, "NASA to Map the Surface of an Asteroid" July 25, 2016 https://www.nasa.gov/feature/goddard/2016/nasa-to-map-the-surface-of-an-asteroid (accessed December 23, 2017).

364. Chris Cooper, Everything You Need to Know About the Universe (San Diego, CA: Thunder Bay Books, 2011) p. 96.

365. Keith Cooper, "Villain in Disguise: Jupiter's Role in Impacts on

Earth", Astrobiology Magazine, March 15, 2012 https://www.space.com/14919-jupiter-comet-impacts-earth.html (accessed December 23, 2017).

366. NASA, "Asteroids in Depth" https://solarsystem.nasa.gov/planets/asteroids/indepth (accessed December 23, 2017).

367. NASA, "Full View of Asteroid Vesta" Sept. 29, 2013 https://www.nasa.gov/mission_pages/dawn/multimedia/pia15678.html (accessed December 23, 2017).

368. NASA, "Hubble Reveals Observable Universe Contains 10 Times More Galaxies Than Previously Thought", Oct. 13, 2016 https://www.nasa.gov/feature/goddard/2016/hubble-reveals-observable-universe-contains-10-times-more-galaxies-than-previously-thought (accessed December 23, 2017).

Christopher J. Conselice, Aaron Wilkinson, Kenneth Duncan, Alice Mortlock, "The Evolution of Galaxy Number Density at Z < 8 And Its Implications", University of Nottingham, School of Physics & Astronomy, https://arxiv.org/pdf/1607.03909v2.pdf (accessed December 23, 2017).

Maria Temming, "Galaxies in the Universe: How Many are There? | Sky and Telescope, July 18, 2014 http://www.skyandtelescope.com/astronomy-resources/how-many-galaxies/ (accessed December 23, 2017).

369. EarthSky, "How long to orbit Milky Way's center?", November 28, 2016 http://earthsky.org/astronomy-essentials/milky-way-rotation (accessed December 27, 2017).

370. Maggie Masetti, "How Many Stars in the Milky Way?" Astrophysics Science Division at NASA's Goddard Space Flight Center, July 22, 2015 https://asd.gsfc.nasa.gov/blueshift/index.php/2015/07/22/how-many-stars-in-the-milky-way/ (accessed December 23, 2017).

Maria Temming, "How Many Stars are There in the Universe?", July 15, 2014 http://www.skyandtelescope.com/astronomy-resources/how-many-stars-are-there/ (accessed December 24, 2017).

371. NASA, "How many exoplanets has Kepler discovered?" Dec. 22, 2017, https://www.nasa.gov/kepler/discoveries (accessed December 27, 2017).

372. Ibid.

373. NASA, "Kepler Habitable Zone Planets", June 19, 2017, https://www.nasa.gov/<INSERT>Image-feature/ames/kepler/kepler-habitable-zone-planets (accessed December 27, 2017).

374. NASA, "The Milky Way Contains at Least 100 Billion Planets According to Survey" Jan 11, 2012 http://hubblesite.org/news_release/news/2012-07/1-planets (accessed December 27, 2017).

375. Ibid.

376. Maria Temming, "How Many Stars are There in the Universe?", July 15, 2014 http://www.skyandtelescope.com/astronomy-resources/how-many-stars-are-there/ (accessed December 24, 2017).

377. Deborah Byrd, "How long to travel to Alpha Centauri?", EarthSky, May 16, 2017, http://earthsky.org/space/alpha-centauri-travel-time (accessed

December 27, 2017).

378. Michael A. Strauss, Welcome to the Universe: An Astrophysical Tour, (Princeton, NJ: Princeton University Press, 2016), p. 185.

379. EarthSky, "Where is the Milky Way?", March 3, 2014, http://earthsky.org/space/galaxy-universe-location (accessed November 1, 2017).

380. Leah Crane, "Galaxy supercluster is one of the biggest things in the universe", New Scientist, 14 July 2017, https://www.newscientist.com/article/2140847-galaxy-supercluster-is-one-of-the-biggest-things-in-the-universe/ (accessed November 2, 2017).

381. Brian Koberlein, interview with the author. December 15, 2017.

382. John Bochanski, interview with the author. December 11, 2017.

CHAPTER 9

384. Lee F. Bennett, ed., Proceedings of the Indiana Academy of Science (Fort Wayne, Indiana: Fort Wayne Printing Company, 1916), p. 83.

384. Bob Brier, Egyptian Mummies: Unraveling the Secrets of an Ancient Art (New York: Harper Perennial, 1996), pp. 61–62.

385. Michael S. Sweeney, Brain: The Complete Mind: How It Develops, How It Works, and How to Keep It Sharp. (Washington, DC: National Geographic, 2009), p. 3.

386. Ananya Mandal, "Semen and Culture," News Medical, May 27, 2015, http://www.news-medical.net/health/Semen-and-Culture.aspx (accessed May 26, 2015). And, Denis Noble, Dario DiFrancesco, and Diego Zancani, "Leonardo da Vinci and the Origin of Semen," Royal Society, Notes and Records, August 20, 2014, http://rsnr.royalsocietypublishing.org/content/early/2014/08/14/rsnr.2014.0021#xref-fn-7-1 (accessed May 27, 2015).

387. Olivia Rudgard, "Belief in witchcraft and demonic possession linked to 1,500 child abuse cases," The Telegraph, 24 November 2017, http://www.telegraph.co.uk/news/2017/11/24/belief-witchcraft-demonic-possession-linked-1500-child-abuse/ (accessed December 13, 2017).

Erin Hale, "Demonic possession in Laos - is it real, or a pretext for village chiefs to banish troublemakers and nonconformists?," Post Magazine, July 22, 2017, http://www.scmp.com/magazines/post-magazine/long-reads/article/2103274/demonic-possession-laos-it-real-or-pretext (accessed December 13, 2017).

388. Carey K. Morewedge, Haewon Yoon, Irene Scopelliti, Carl W. Symborski, James H. Korris, and Karim S. Kassam, "Debiasing Decisions: Improved Decision Making With a Single Training Intervention," Policy Insights from the Behavioral and Brain Sciences 2015, Vol. 2(1) 129 –140 DOI: 10.1177/2372732215600886 (accessed December 21, 2017).

389. Gary Marcus, Kluge: The Haphazard Construction of the Human Mind (New York: Houghton Mifflin, 2008), p. 6.

390. Ian Tattersall, Becoming Human: Evolution and Human Uniqueness (New York: Harcourt Brace, 1998), pp. 72–73.

391. Ibid, pp. 72-73.

392. Jeremy Hsu, "How Much Power Does The Human Brain Require To Operate?," Popular Science, November 6, 2009, https://www.popsci.com/technology/article/2009-11/neuron-computer-chips-could-overcome-power-limitations-digital (accessed January 11, 2018).

393. Ponce de León MS, Golovanova L, Doronichev V, et al. Neanderthal brain size at birth provides insights into the evolution of human life history. Proceedings of the National Academy of Sciences of the United States of America. 2008;105(37):13764-13768. doi:10.1073/pnas.0803917105.

394. Herculano-Houzel S, Avelino-de-Souza K, Neves K, et al. The elephant brain in numbers. Frontiers in Neuroanatomy. 2014;8:46. doi:10.3389/fnana.2014.00046.

395. F. A. Azevedo, L. R. Carvalho, L. T. Grinberg, J. M. Farfel, R. E. Ferretti, R. E. Leite, W. Jacob Filho, R. Lent, and S. Herculano-Houzel, "Equal Numbers of Neuronal and Nonneuronal Cells Make the Human Brain an Isometrically Scaled-Up Primate Brain," Journal of Comparative Neurology 513, no. 5 (April 10, 2009): 532–41, http://www.ncbi.nlm.nih.gov/pubmed?Db=pubmed&Cmd=ShowDetailView&TermToSearch=19226510 (accessed January 19, 2015).

396. Michael Balter, "Closer Look at Einstein's Brain" Science Apr. 17, 2009   http://www.sciencemag.org/news/2009/04/closer-look-einsteins-brain (accessed December 21, 2017).

397. Christof Koch, "Does Size Matter—for Brains?," Scientific American, January 1, 2016, https://www.scientificamerican.com/article/does-size-matter-for-brains/ (accessed December 14, 2017).

398. Seid MA1, Castillo A, Wcislo WT, "The allometry of brain miniaturization in ants,"
Brain Behavior and Evolution, 2011;77(1):5-13. doi: 10.1159/000322530. (accessed December 14, 2017).

399. Ray Kurzweil, How to Create a Mind: The Secret of Human Thought Revealed, (New York, NY: Viking, 2012) p. 172.

400. K. S. Krabbe, A. R. Nielsen, R. Krogh-Madsen, P. Plomgaard, P. Rasmussen, C. Erikstrup, C. P. Fischer, B. Lindegaard, A. M. Petersen, S. Taudorf, N. H. Secher, H. Pilegaard, H. Bruunsgaard, and B. K. Pedersen, "Brain-Derived Neurotrophic Factor (BDNF) and Type 2 Diabetes," Diabetologia 50, no. 2 (February 2007): 431–38.

401. R. Molteni, R. J. Barnard, Z. Ying, C. K. Roberts, and F. GómezPinilla, "A High-Fat, Refined Sugar Diet Reduces Hippocampal Brain-Derived Neurotrophic Factor, Neuronal Plasticity, and Learning," Neuroscience 112,

no. 4 (2002):803–14, http://www.ncbi.nlm.nih.gov/pubmed/12088740 (accessed December 4, 2015).

402. Lucia Kerti, A. Veronica Witte, Angela Winkler, Ulrike Grittner, Dan Rujescu, and Agnes Flöel, "Higher Glucose Levels Associated with Lower Memory and Reduced Hippocampal Microstructure," Neurology 81, no. 20 (November 12, 2013): 1746–752, http://www.neurology.org/content/81/20/1746 (accessed December 19, 2017).

403. "Drinking Sugar-Sweetened Beverages during adolescence Impairs Memory, Animal Study Suggests," Science Daily, July 29, 2014, http://www .sciencedaily.com/releases/2014/07/140729224906.htm (accessed January 2, 2015).

404. Quoted in Stuart Wolpert, "Scientists Learn How What You Eat Affects Your Brain—and Those of Your Kids," UCLA Newsroom, July 9, 2008, http:// newsroom.ucla.edu/releases/scientists-learn-how-food-affects-52668 (accessed December 20, 2017).

405. Martha Clare Morris, Yamin Wang, Lisa L. Barnes, David A. Bennett, et al. "Nutrients and bioactives in green leafy vegetables and cognitive decline" Neurology December 20, 2017, DOI: https://doi.org/10.1212/WNL.0000000000004815 (accessed December 21, 2017).

Federation of American Societies for Experimental Biology (FASEB), "Eating Green Leafy Vegetables Keeps Mental Abilities Sharp," ScienceDaily, March 30, 2015, www.sciencedaily.com/releases/2015/03/150330112227.htm (accessed December 19, 2017).

406. American Chemical Society. "Blueberries, the well-known 'super fruit,' could help fight Alzheimer's." ScienceDaily. ScienceDaily, 14 March 2016. www.sciencedaily.com/releases/2016/03/160314084821.htm (accessed December 21, 2017).

Marshall G. Miller, Barbara Shukitt-Hale. Berry Fruit Enhances Beneficial Signaling in the Brain. *Journal of Agricultural and Food Chemistry*, 2012; 120203155528007 DOI: 10.1021/jf2036033 (accessed December 21, 2017).

Joanna L. Bowtell, Zainie Aboo-Bakkar, Myra Conway, Anna-Lynne R. Adlam, Jonathan Fulford. Enhanced task related brain activation and resting perfusion in healthy older adults after chronic blueberry supplementation. *Applied Physiology, Nutrition, and Metabolism*, 2017; DOI: 10.1139/apnm-2016-0550 (accessed December 21, 2017).

407. Marta K. Zamroziewicz, M. Tanveer Talukdar, Chris E. Zwilling, Aron K. Barbey. Nutritional status, brain network organization, and general intelligence. NeuroImage, 2017; 161: 241 DOI: 10.1016/j.neuroimage.2017.08.043 (accessed December 21, 2017).

408. Valerie Ross, "Numbers: The Nervous System, From 268-MPH Signals to Trillions of Synapses," Discover, May 15, 2011, http://discovermagazine.com/2011/mar/10-numbers-the-nervous-system (accessed January 11, 2018).

409. John J. Ratey, Spark: The Revolutionary New Science of Exercise and

the Brain (New York: Little, Brown, 2013), pp. 35–36.

410. V. S. Ramachandran, The Tell-Tale Brain: A Neuroscientist's Quest for What Makes Us Human (New York: W. W. Norton, 2011), p. 14.

411. Ibid.

412. Ibid, Ratey, pp. 35-36.

413. Peter S. Eriksson, Ekaterina Perfilieva, Thomas Björk-Eriksson, Ann-Marie Alborn, Claes Nordborg, Daniel A. Peterson, and Fred H. Gage, "Neurogenesis in the Adult Human Hippocampus," Nature Medicine 4, no. 11 (November 1998).

414. Kirk I. Erickson, Ruchika S. Prakash, Michelle W. Voss, Laura Chaddock, Liang Hu, Katherine S. Morris, Siobhan M. White, Thomas R. Wójcicki, Edward McAuley, Arthur F. Kramer. Aerobic fitness is associated with hippocampal volume in elderly humans. Hippocampus, 2009; NA DOI: 10.1002/hipo.20547 (accessed December 21, 2017).

M. Kodali, T. Megahed, V. Mishra, B. Shuai, B. Hattiangady, A. K. Shetty. Voluntary Running Exercise-Mediated Enhanced Neurogenesis Does Not Obliterate Retrograde Spatial Memory. Journal of Neuroscience, 2016; 36 (31): 8112 DOI: 10.1523/JNEUROSCI.0766-16.2016 (accessed December 21, 2017).

415. John J. Ratey, Spark: The Revolutionary New Science of Exercise and the Brain (New York: Little, Brown, 2013), p. 242.

416. Brenda Patoine, "Move Your Feet, Grow New Neurons?" Dana Foundation, May 2007, http://www.dana.org/Publications/Brainwork/Details.aspx?id=43678 (accessed December 20, 2017).

417. Ibid.

418. K. A. Barbour, T. M. Edenfield, and J. A. Blumenthal, "Exercise as a Treatment for Depression and Other Psychiatric Disorders: A Review," Journal of Psychosomatic Medicine 27, no. 6 (November/December 2007): 359–67, doi:10.1097/01.HCR.0000300262.69645.95.

419. Ratey, Spark, pp. 122–23, 135.

420. Dean Buonomano, Brain Bugs: How the Brain's Flaws Shape Our Lives (New York: W. W. Norton, 2011), p. 16.

421. D. A. Pardini, A. Raine, K. Erickson, and R. Loeber, "Lower Amygdala Volume in Men Is Associated with Childhood Aggression, Early Psychopathic Traits, and Future Violence," Biological Psychiatry 75, no. 1 (January 2014): 73–80.

422. A. Abigail Marsh, Sarah A. Stoycos, Kristin M. Brethel-Haurwitz, Paul Robinson, John W. VanMeter, and Elise M. Cardinale, "Neural and Cognitive Characteristics of Extraordinary Altruists," PNAS 111, no. 42 (October 21, 2014): 15036–41.

423. M. S. Christian and A. P. Ellis, "Examining the Effects of Sleep Deprivation on Workplace Deviance: A Self-Regulatory Perspective," Academy of Management Journal 54, no. 5 (2011): 913–34.

424. Katie Moisse, "5 Health Hazards Linked to Lack of Sleep," June 11,

2012, http://abcnews.go.com/Health/Sleep/health-hazards-linked-lack-sleep/story?id=16524313 (accessed December 20, 2017).

Fisher Center for Alzheimer's Research Foundation, "Poor Sleep May Be Linked to Alzheimer's Disease," http://www.alzinfo.org/articles/poor-sleep-may-be-linked-to-alzheimers-disease/ (accessed December 20, 2017).

425. Paul E. Bendheim, The Brain Training Revolution: A Proven Workout for Healthy Brain Aging (Naperville, IL: Sourcebooks, 2009), p. 256.

426. Paul E. Bendheim, The Brain Training Revolution: A Proven Workout for Healthy Brain Aging (Naperville, IL: Sourcebooks, 2009) p. 266.

427. Sara Fattinger, Toon T. de Beukelaar, Kathy L. Ruddy, Carina Volk, Natalie C. Heyse, Joshua A. Herbst, Richard H. R. Hahnloser, Nicole Wenderoth, Reto Huber. Deep sleep maintains learning efficiency of the human brain. Nature Communications, 2017; 8: 15405 DOI: 10.1038/ncomms15405 (accessed December 20, 2017).

428. Majid Fotuhi, David Do, and Clifford Jack, "Modifiable Factors That Alter the Size of the Hippocampus with Ageing," Nature Reviews: Neurology, advance online publication, March 13 2012, pp. 1–2, doi:10.1038/nrneurol.2012.27.

429. Ibid., p. 2.

430. F. A. Azevedo, L. R. Carvalho, L. T. Grinberg, J. M. Farfel, R. E. Ferretti, R. E. Leite, W. Jacob Filho, R. Lent, and S. Herculano-Houzel, "Equal Numbers of Neuronal and Nonneuronal Cells Make the Human Brain an Isometrically Scaled-Up Primate Brain," Journal of Comparative Neurology 513, no. 5 (April 10, 2009): 532–41, http://www.ncbi.nlm.nih.gov/pubmed?Db=pubmed&Cmd=ShowDetailView&TermToSearch=19226510 (accessed December 19, 2017).

431. Paul E. Bendheim, The Brain Training Revolution: A Proven Workout for Healthy Brain Aging (Naperville, IL: Sourcebooks, 2009), p. 9.

432. Floyd E. Bloom, ed., Best of the Brain from Scientific American: Mind, Matter, and Tomorrow's Brain (New York: Dana, 2007), p. 92.

433. Ibid., pp. 92–93.

434. Ibid., p. 93.

435. Jim Bower, interview with the author, January 28, 2015.

436. Ferris Jabr, "Know Your Neurons: What Is the Ratio of Glia to Neurons in the Brain?" Brain Waves (Scientific American blog), June 13, 2012, http://blogs.scientificamerican.com/brainwaves/know-your-neurons-what-is-the-ratio-of-glia-to-neurons-in-the-brain/ (accessed December 19, 2017).

437. Douglas Fields, "Neuroscience: Map the Other Brain," Nature (September 4, 2013), http://www.nature.com/news/neuroscience-map-the-other-brain-1.13654 (accessed December 1, 2014).

438. Ibid.

439. The Brain Made Simple, "Cerebral Cortex," http://brainmadesimple.com/cortex-and-lobes-of-the-brain.html (accessed January 12, 2018).

440. Brenda Patoine, "The Prefrontal Cortex and Frontal Lobe Disorders: An Interview with Jordan Grafman, PhD," Dana Foundation, January, 2006, http://www.dana.org/Publications/ReportDetails.aspx?id=44153 (accessed December 19, 2017).

441. Ezequiel Morsella, et al, "Homing in on consciousness in the nervous system: An action-based synthesis," Behavioral and Brain Sciences, Volume 39, 2016, e168, https://doi.org/10.1017/S0140525X15000643 (accessed December 14, 2017).

442. Beth Tagawa, "Consciousness has less control than believed, according to new theory," SF State News, June 2015 (accessed December 14, 2017).

443. Ibid.

444. Raymond S. Nickerson, "Confirmation Bias: A Ubiquitous Phenomenon in Many Guises," Review of General Psychology, 1998, Vol. 2, No. 2, pp. 175-220.

445. Hank Davis, Caveman Logic: The Persistence of Primitive Thinking in a Modern World (Amherst, NY: Prometheus Books, 2009), pp. 183–84.

446. Daniel Sokal, "Inside the Mind of the Doctor," BBC, May 9, 2007, http://news.bbc.co.uk/2/hi/health/6610719.stm (accessed December 15, 2017)

447. Ibid.

448. Gregory N. Bratman, et al. "The benefits of nature experience: Improved affect and cognition" Landscape and Urban Planning Volume 138, June 2015, Pages 41-50 https://doi.org/10.1016/j.landurbplan.2015.02.005 (accessed December 20, 2017).

Roger S.Ulrich, Robert F.Simons†Barbara, et al. "Stress recovery during exposure to natural and urban environments," *Journal of Environmental Psychology* Volume 11, Issue 3, September 1991, Pages 201-230 https://doi.org/10.1016/S0272-4944(05)80184-7 (accessed December 20, 2017).

449. Irene Scopelliti, Carey K. Morewedge, Erin McCormick, H. Lauren Min, Sophie Lebrecht, and Karim S. Kassam, "Bias Blind Spot: Structure, Measurement, and Consequences," Management Science, 201561:10 , 2468-2486 https://pubsonline.informs.org/doi/abs/10.1287/mnsc.2014.2096 (accessed January 12, 2018).

450. Stanley Milgram, "Behavioral Study of obedience," The Journal of Abnormal and Social Psychology. 67 (4). July 1963.

Jennifer L. Devenport, Ronald P. Fisher, "The effect of authority and social influence on eyewitness suggestibility and person recognition," *Journal of Police and Criminal Psychology*, March 1996, Volume 11, Issue 1, pp 35–40.

451. ABC News, "Restaurant Shift Turns into Nightmare," November 10, 2005, http://abcnews.go.com/Primetime/story?id=1297922&page=1&singlePage=true (accessed December 21, 2017).

452. Stanley Milgram, Obedience to Authority: An Experimental View (New York, NY: Harper and Row), 1974.

453. Ryan McElhany, "The Effects of Anchoring Bias on Human Behavior,"

Thought Hub, May 23, 2016 https://www.sagu.edu/thoughthub/the-affects-of-anchoring-bias-on-human-behavior (accessed January 12, 2018).

454. AdrianFurnham, Hua ChuBoo, "A literature review of the anchoring effect," The Journal of Socio-Economics Volume 40, Issue 1, February 2011, Pages 35-42 https://doi.org/10.1016/j.socec.2010.10.008 (accessed December 21, 2017).

A. Tversky and D. Kahneman, "Judgment under Uncertainty: Heuristics and Biases," Science 185, no. 4157 (1974): 1124–31, doi:10.1126/science.185.4157.1124.

455. Keith E. Stanovich, Richard West, "On the Relative Independence of Thinking Biases and Cognitive Ability" Journal of Personality and Social Psychology, 2008, Vol. 94, No. 4, 672–695 DOI: 10.1037/0022-3514.94.4.672 (accessed December 21, 2017).

456. Keith E. Stanovich, What Intelligence Tests Miss: The Psychology of Rational Thought (Connecticut: Yale University Press, 2010).

Keith E. Stanovich, Richard F. West, Maggie E. Toplak, The Rationality Quotient: Toward a Test of Rational Thinking (Cambridge, MAMIT Press, 2018).

457. Ibid. Morewedge.

458. D. J. Simons, C. F. Chabris, "What People Believe about How Memory Works: A Representative Survey of the U.S. Population," PLoS ONE 6, no. 8 (August 3, 2011): e22757, doi:10.1371/journal.pone.0022757, http://journals.plos.org/plosone/article?id=10.1371/journal.pone.0022757 (accessed January 1, 2018).

459. Ibid.

460. Dean Buonomano, Brain Bugs: How the Brain's Flaws Shape Our Lives (New York: W. W. Norton, 2011), p. 68.

461. Elizabeth Loftus, "The Reality of Repressed Memories," American Psychologist 48, no. 5 (May 1993): p. 524.

462. Elizabeth Loftus, "Elizabeth Loftus: The Fiction of Memory," 17:36, TED talk filmed June 2013, https://www.ted.com/talks/elizabeth_loftus_the _ fiction_of_memory#t-316982 (accessed January 27, 2018).

CHAPTER 10

464. Evolution (PBS), "The Current Mass Extinction," http://www.pbs.org/wgbh/evolution/library/03/2/l_032_04.html, (accessed January 19, 2018).

464. The Project for Future Human Flourishing, https://www.xrisksinstitute.com/

465. Phil Torres, interview with the author, January 17, 2018.

466. NASA's Planetary Defense Office, https://www.nasa.gov/plane-

tarydefense/overview

B612 Foundation, https://b612foundation.org/

467. Donald Lowe, interview with the author, January 25, 2018.

468. David Quammen, ed., Charles Darwin: On the Origin of Species: The Illustrated Edition (New York, Sterling Signature, 2011), p. 513.

469. Cameron M. Smith ,"Estimation of a genetically viable population for multigenerational interstellar voyaging: Review and data for project Hyperion," Acta Astronautica, Volume 97, April–May 2014, pp. 16-29, https://doi.org/10.1016/j.actaastro.2013.12.013 (accessed January 12, 2018).

470. Michael R. Rampino, Stanley H. Ambrose, "Volcanic winter in the Garden of Eden: The Toba supereruption and the late Pleistocene human population crash," Special Paper of the Geological Society of America, V. 342, 2000, pp. 71-82, https://nyuscholars.nyu.edu/en/publications/volcanic-winter-in-the-garden-of-eden-the-toba-supereruption-and- (accessed January 12, 2018).

471. CBS, "Does supervolcano under Yellowstone have planet-killing potential?" October 13, 2017https://www.cbsnews.com/news/yellowstone-national-park-supervolcano-caldera/ (accessed January 14, 2018).

472. Jonathan, Rougiera, et al., "The global magnitude–frequency relationship for large explosive volcanic eruptions," Earth and Planetary Science Letters, Volume 482, 15 January 2018, Pages 621-629 doi.org/10.1016/j.epsl.2017.11.015 (accessed January 14, 2018).

473. NASA/Jet Propulsion Laboratory, "The Probability of Collisions with Earth," https://www2.jpl.nasa.gov/sl9/back2.html (accessed January 16, 2018) .

474. Daniel Enberg, "How Big Would A Meteorite Have To Be To Wipe Out All Human Life?," Popular Science, February 26, 2015, https://www.popsci.com/how-big-would-meteorite-have-be-wipe-out-all-human-life (accessed December 26, 2017).

475. Norman H. Sleep, Donald R. Lowe, "Physics of crustal fracturing and chert dike formation triggered by asteroid impact, 3.26 Ga, Barberton greenstone belt, South Africa," Geochemistry, Geophysics, Geosystem, Volume 15, Issue 4, April 2014, 1054–1070. http://onlinelibrary.wiley.com/doi/10.1002/2014GC005229/full (accessed January 23, 2018).

476. NASA, "Asteroids in Depth" https://solarsystem.nasa.gov/planets/asteroids/indepth (accessed December 23, 2017).

477. Planetary Defense (NASA),"Planetary Defense Frequently Asked Questions," https://www.nasa.gov/planetarydefense/faq (accessed January 23, 2018).

478. NASA/Jet Propulsion Laboratory, "The Probability of Collisions with Earth," https://www2.jpl.nasa.gov/sl9/back2.html (accessed January 16, 2018) .

479. NASA, "The Spaceguard Survey: NASA International Near-Earth-Object Detection Workshop," January 25, 1992, https://ntrs.nasa.gov/archive/nasa/casi.ntrs.nasa.gov/19920025001.pdf (accessed January 16, 2018), p. 2.

480. Ibid, pp. 10-11.

481. Planetary Defense (NASA),"Planetary Defense Frequently Asked Questions," https://www.nasa.gov/planetarydefense/faq (accessed January 23, 2018).

482. Ibid.

483. "The world in 2076: Machines outsmart us but we're still on top" New Scientist 232, no. 3100, https://www.newscientist.com/article/mg23231000-600-the-world-in-2076-machines-outsmart-us-but-were-still-on-top/ (accessed December 4, 2017).

484. "NGA 2017 Summer Meeting—Introducing the New Chair's Initiative 'Ahead of the Curve,'" YouTube video, 1:26:50, streamed live July 15, 2017, posted by National Governors Association, https://www.youtube.com/watch?v=2C-A797y8dA (accessed December 26, 2017).

485. Rory Cellan-Jones, "Stephen Hawking Warns Artificial Intelligence Could End Mankind," BBC News, December 2, 2014, http://www.bbc.com/news/technology-30290540 (accessed December 9, 2017)

486. Nick Bostrom, Superintelligence: Paths, Dangers, Strategies (Oxford: Oxford Press, 2014), p. 319.

487. Ariel Conn, "Can We Properly Prepare for the Risks of Superintelligent AI?" Future of Life Institute, March 23, 2017, https://futureoflife.org/2017/03/23/ai-risks-principle/ (accessed January 14, 2018).

488. Guy P. Harrison, Think Before You Like: Social Media's Effect on the Brain and the Tools You Need to Navigate Your Newsfeed (Amherst, NY, Prometheus Books, 2017).

489. Josh Constine, "Facebook Is Building Brain-Computer Interfaces for Typing and Skin-Hearing," TechCrunch, April 19, 2017, https://techcrunch.com/2017/04/19/facebook-brain-interface/ (accessed December 8, 2017).

490. Cade Metz, "Elon Musk Isn't the Only One Trying to Computerize Your Brain," Wired, March 31, 2017, https://www.wired.com/2017/03/elonmusks -neural-lace-really-look-like (accessed April 28, 2017).

491. Ethan Siegel, interview with the author, December 12, 2017.

492. Future of Life Institute, "Accidental Nuclear War: A Timeline of Close Calls" https://futureoflife.org/background/nuclear-close-calls-a-timeline/ (accessed January 14, 2018).

Robert Krulwich, "You (and Almost Everyone You Know) Owe Your Life to This Man" National Geographic, March 25, 2016, https://news.national-geographic.com/2016/03/you-and-almost-everyone-you-know-owe-your-life-to-this-man/# (accessed January 14, 2018).

493. Brooking, "50 Facts About U.S. Nuclear Weapons Today," April 28, 2014, https://www.brookings.edu/research/50-facts-about-u-s-nuclear-weapons-today/ (accessed December 31, 2017).

494. Ibid.

495. Hans M. Kristensen, Robert S. Norris, "Status of World Nuclear Forces," Federation of American Scientists," https://fas.org/issues/nuclear-

weapons/status-world-nuclear-forces/ (accessed January 1, 2018).

496. Michael J. Mills, Owen B. Toon, et al, "Multidecadal global cooling and unprecedented ozone loss following a regional nuclear conflict," Earth's Future, Volume 2, Issue 4, April 2014, pp. 161–176.

497. Ibid.

498. Jennifer Viegas, "Human extinction: How could it happen?," NBC News, November 11, 2009 http://www.nbcnews.com/id/33859415/ns/technology_and_science-science/t/human-extinction-how-could-it-happen/#. Wl46GkxFzIV (accessed January 16, 2018).

Lopes, T. P., Chermack, T. J., Demers, D., Kasshanna, B., Payne, T., Kari, M. " Human Extinction Scenario Frameworks," Futures, 41(10), 2009, 731-737.

499. Steven Pinker, Better Angels of Our Nature: Why Violence Has Declined, (New York, NY: Viking, 2011).

500. Oliver Milman, "Rate of environmental degradation puts life on Earth at risk, say scientists," The Guardian, January 15, 2015, https://www.theguardian.com/environment/2015/jan/15/rate-of-environmental-degradation-puts-life-on-earth-at-risk-say-scientists (accessed January 26, 2018).

Antonio Savoia, "Global inequality is on the rise – but at vastly different rates across the world," The Conversation, December 14, 2017, https://theconversation.com/global-inequality-is-on-the-rise-but-at-vastly-different-rates-across-the-world-88976 (accessed January 26, 2018).

501. Kelsey D. Atherton, "The Pentagon's new drone swarm heralds a future of autonomous war machines," Popular Science, January 10, 2017, https://www.popsci.com/pentagon-drone-swarm-autonomous-war-machines (accessed January 26, 2018).

Paul Scharre, "Why We Must Not Build Automated Weapons of War," Time, September 25, 2017, http://time.com/4948633/robots-artificial-intelligence-war/ (accessed January 26, 2018).

502. Stephen Hawking, "This is the most dangerous time for our planet," The Guardian, December 1, 2016, https://www.theguardian.com/commentisfree/2016/dec/01/stephen-hawking-dangerous-time-planet-inequality (accessed January 28, 2018).

503. Guy P. Harrison, "Fighting the Bomb," Caymanian Compass, March 28, 2003, pp. A20-A21.

504. Guy P. Harrison, "The Man Who Won World War III," Caymanian Compass, March 14, 2002, p. 14.

505. Chris Impey, How It Ends: From You To the Universe, (New York, NY: W.W. Norton, 2010), p. 178.

506. Phil Plait, Death From the Skies! The Science Behind the End of the World, (New York, NY: Penguin, 2009), p. 119.

507. Ibid, p. 121.

508. Ibid, p. 121.

509. Impey, p. 180.

510. Philip Plait, *Death From the Skies: The Science Behind the End of the World*, (New York, NY: Penguin Books) pp. 198-208.

Chris Impey, *How IT Ends: From You to the Universe*, (New York, NY: Norton, 2011) pp. 187-191.

511. Ibid, Plait, p. 205.

512. Joshua Sokol, "Mystery bright spots could be first glimpse of another universe," *New Scientist*, 28 October 2015, https://www.newscientist.com/article/2063204-mystery-bright-spots-could-be-first-glimpse-of-another-universe/ (accessed January 1, 2018).

513. Camilo Mora, Derek P. Tittensor, Sina Adl, Alastair G. B. Simpson, Boris Worm, "How Many Species Are There on Earth and in the Ocean?" PLOS Biology, August 23, 2011 https://doi.org/10.1371/journal.pbio.1001127 (accessed November 9, 2017).

514. Nathan Collins, "Stanford study indicates that more than 99 percent of the microbes inside us are unknown to science," Stanford News, August 22, 2017, https://news.stanford.edu/2017/08/22/nearly-microbes-inside-us-unknown-science/ (accessed January 9, 2018).

515. Jennifer Welsh, "Tools May Have Been First Money," Live Science, February 29, 2012, https://www.livescience.com/18751-hand-axe-tools-money.html (accessed January 8, 2018).

1. 17 purpose

2. 63 microbiome

3. 84 too much sitting
   walking